MICROCOMPUTERS AND LABORATORY INSTRUMENTATION

MICROCOMPUTERS AND LABORATORY INSTRUMENTATION

David J. Malcolme-Lawes

King's College London
London, England

PLENUM PRESS • NEW YORK AND LONDON

Library of Congress Cataloging in Publication Data

Malcolme-Lawes, D. J.
   Microcomputers and laboratory instrumentation.

   Bibliography: p.
   Includes index.
   1. Physical instruments—Data processing. 2. Physical laboratories—Data process-
ing. 3. Microcomputers. I. Title.
QC53.M27   1984   530'.7'02854   84-3328
ISBN 0-306-41668-9

© 1984 Plenum Press, New York
A Division of Plenum Publishing Corporation
233 Spring Street, New York, N.Y. 10013

Printed in the United States of America

To Louisa and James

PREFACE

The invention of the microcomputer in the mid-1970s and its subsequent low-cost proliferation has opened up a new world for the laboratory scientist. Tedious data collection can now be automated relatively cheaply and with an enormous increase in reliability. New techniques of measurement are accessible with the "intelligent" instrumentation made possible by these programmable devices, and the ease of use of even standard measurement techniques may be improved by the data processing capabilities of the humblest micro. The latest items of commercial laboratory instrumentation are invariably "computer controlled", although this is more likely to mean that a microprocessor is involved than that a versatile microcomputer is provided along with the instrument.

It is clear that all scientists of the future will need some knowledge of computers, if only to aid them in mastering the button pushing associated with gleaming new instruments. However, to be able to exploit this newly accessible computing power to the full the practising laboratory scientist must gain sufficient understanding to utilise the communication channels between apparatus on the laboratory bench and program within the computer. This book attempts to provide an introduction to those communication channels in a manner which is understandable for scientists who do not specialise in electronics or computers.

The contents are based on courses given to undergraduate and postgraduate science students at King's College London. The objective of those courses was to provide students with an understanding of how modern microcomputers can communicate with laboratory apparatus for measurement and control purposes. It was not expected that all the students would have to design and build interfaces to achieve their ends, but rather that they should understand the principles on which interfaces operate and the capabilities and limitations of practical devices, so that they could design experiments in their own fields with a foundation knowledge of how a microcomputer could be employed.

The courses were closely associated with practical experience gained on microcomputers and a variety of items of standard laboratory instrumentation. Of course that element is not included in the present text, but the fact remains that this book is intended to be of assistance to the practical scientist. While not designed as a do-it-yourself guide to building particular electronic circuits, a number of interfacing circuits are discussed in some detail and the readers may well be able to develop these to suit their own needs. The circuits described are derived from the author's own experience, which is limited to systems associated with PET, Apple, BBC and Spectrum microcomputers. The devices available for use in signal handling continue to increase rapidly. Even over the last year a number of new devices have appeared (particularly opto-isolated devices and LSI circuits), which would be useful for a number of the tasks discussed. However, the principles of communication between microcomputers and laboratory systems are not changing quite so rapidly, and it is hoped that the examples will be found useful.

I wish to express my thanks to Drs. Allwood, Blatchford and Overill of King's College London for their helpful advice and criticism during the preparation of this manuscript, to the research students who have suffered changes in the apparatus while circuits were tested, and to my family for their patience during my long sessions with the WORDPET word processor.

King's College London   D. J. Malcolme-Lawes
October 1983

CONTENTS

CHAPTER 1

INTRODUCTION

One of the consequences of a major change in technology is that a large number of familiar ideas, procedures and equipment quickly become replaced by less familiar devices and ways of doing things. The world has recently embarked on one such revolutionary upheaval and, although it will be many years before the full impact of computers is felt, it is now that we should prepare ourselves to understand the nature of the massive changes which lie ahead. The scale of the technological changes to come will undoubtedly be enormous, just as was the case during the first industrial revolution, and go well beyond the perceptions of those who think that "it" has already happened. These changes will eventually effect almost every aspect of our lives, from education to leisure, from housekeeping to heavy industry. In this book we are concerned with the beginings of this change in one small area of activity - the laboratory, and in particular with the electronic instrumentation used in the laboratory for measurement and control purposes.

Let us start by making quite clear the aim of this book. The objective is to provide scientists who do not specialise in electronics or computers with an introduction to the basic aspects of the use of low cost, mass produced microcomputers for communicating with and controlling experimental equipment. It is not part of the objective to discuss robotics, the mathematical procedures of data handling or sophisticated electronic signal processing techniques. Neither is the use of microprocessors in laboratory instrumentation to be discussed and it is not assumed that the reader has any special knowledge of these devices - other than that a microprocessor is one of the component parts of a microcomputer. In this chapter we discuss briefly some general aspects of microcomputers and their relation to laboratory instrumentation, and outline the level of knowledge required to appreciate subsequent chapters. In chapter 2 we examine the basics of the electrical signals commonly encountered in laboratory systems, and in chapters 3 and 4 we discuss the major elements of the electronic circuits used for handling analog and digital signals respectively. In chapter 5 we look at the modern microcomputer, some of its peripherals which are of value in the laboratory, and a few of the details of its internal organisation and

Table 1.1 Examples of Measurement and Control functions
         of commonly used laboratory instruments

Instrument	Measured property
thermometer	temperature
manometer	pressure
photometer	light intenstity
pH meter	hydrogen ion activity
GM counter	radioactivity
Multimeter	voltage, current, resistance
clock	time

Instrument	Controlled property
thermostat	temperature
manostat	pressure
potentiostat	applied voltage
monochromator	wavelength transmitted
timer	time interval
flostat	flow rate

function - where these are related to communication with external
equipment. Chapters 6 and 7 are devoted to some of the techniques
which can be used for communication between a microcomputer and other
items of analog or digital signal handling equipment - a subject known
almost universally as interfacing. In chapter 8 we consider how to
approach the problem of designing an instrumentation system which uses
a microcomputer as its control centre, looking at both the hardware
and software aspects of the problem.

It must be pointed out that the subject matter of the book is
intended to overlap with a number of traditional fields, and that the
topics discussed are covered solely for their relevance to the
application of microcomputers in practical laboratories. The coverage
can undoubtedly be criticised for its omission of topics necessary for
a wider understanding of electronics or computers, and could equally
be regarded as unnecessary for the man who just wants to connect a
computer to an instrument to collect some data. However, the nature of
laboratories will change as a result of the technological
reorientation that computers are bringing, and in this author's view
many scientists, whatever their specialist field, will require a
quantity of knowledge in these areas.

Table 1.2   Examples of low cost microcomputers

Models	Manufacturer
BBC model A & B	Acorn Computers
Electron	Acorn Computers
Apple II and III	Apple Computers
CBM, PET, VIC & 64	Commodore
Dragon	Dragon
HX-20	Epson
Newbrain	Grundy Business Machines
HP85, HP125	Hewlett Packard
380Z, 480Z	Research Machines
ZX81 and Spectrum	Sinclair Research
Oric	Tangerine Computers
TRS80 and model 3	Tandy

1.1 Laboratory instrumentation and microcomputers

Laboratory instrumentation assists the scientist by enabling him
to make measurements of a wide range of physical, chemical and
biological parameters, and by automating the control of measurements,
processes and recording functions. Examples of just a few of the
measurement and control functions of widely used laboratory
instrumentation are listed in table 1.1.

Modern instrumentation allows sophisticated combinations of many
of the basic measurement and control functions to provide systems
capable of the direct and highly automated measurement of complex
quantities. Examples include systems such as thermal luminescent
dosimeters, automatic liquid scintillation counters, materials testing
equipment, magnetic resonance spectrometers, infra red and ultra
violet absorption spectrometers and gradient elution chromatographs.
However, although well established instrumental techniques have
benefitted from recent developments in electronics, the situation for
laboratories with limited resources and for research groups working in
new areas of measurement and control is less favourable if commercial
systems which meet their specifications are not available.

The dramatic growth in the availability of low cost
microcomputers over the last few years has initiated significant
changes in the development of laboratory instrumentation, not only by
allowing simplification of the operation of many common instruments,
but also by providing the means for the development of new instruments
based on methods of measurement and control complexity which were
relatively inaccessible to the previous generation, such as real-time

Table 1.3   Typical BASIC language instructions

```
 1 REM Remarks for ease of reading

10 INPUT X                :REM input data and store in X
20 LET Y = 2+X/3          :REM calculate Y from expression
30 GOTO 100               :REM control transferred to 100
40 PRINT "Y= "; Y         :REM output message & Y
50 IF X=0 THEN 100        :REM goto 100 if X is 0
70 STOP                   :REM stop execution of program

60 FOR I=1 TO 100         :REM instructions between FOR and
   - - - - - -            :REM NEXT repeated for values of
69 NEXT I                 :REM I from 1 to 100 inclusive

80 GOSUB 5000             :REM control transferred to line
5000 Y=0                  :REM 5000 then instructions obeyed
5999 RETURN               :REM until a RETURN is encountered
```

Fourier transform techniques and diode array detectors for
spectroscopy. Some of the most popular microcomputers, available at
costs ranging from less than $100 to over $2000, are listed in table
1.2. Any of these micros may be used for the moderately high speed
measurement of signals, storing of measurements and data, controlling
electrical and mechanical equipment, performing arithmetical and
logical manipulations, and the displaying or recording of
information.

Microcomputers offer the laboratory scientist a new dimension in
instrumentation because they can be programmed to perform the tasks o
complex electronic circuits, and the program can be modified until it
does precisely what is wanted. Furthermore, when properly designed and
tested, microcomputer based systems can offer fast and reliable
operation over long periods of time and yet can be rapidly
reconfigured to perform a totally different function when necessary.
They offer the laboratory scientist the ability to control
instumentation during long term unattended operation, to automate
routine measurement functions for large numbers of samples, to record
and process large amounts of data, and to present the result of a
measurement based on many or different quantities.

The value of the modern microcomputer lies in the fact that it
can be programmed. Almost anyone with a secondary school education can
master the elements of programming a microcomputer in an afternoon
using the most common language (BASIC). The principal instructions of
the BASIC language are listed in table 1.3, although we are not going
to cover the language further or discuss its normal use here. Indeed

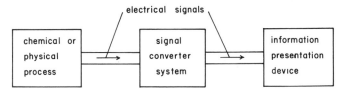

Fig 1.1 The basic components of a conventional measurement system.

we will assume that the reader is familiar with both the appearance of microcomputers and with the elements of BASIC (or any other BASIC-like language, such as FORTRAN) when we come to discuss the operation of micros and communication with them. (There is a large number of books available on the use of BASIC and it is to be highly recommended that the reader who is not already familiar with the language should study one of these before going beyond chapter 4 in this text.)

1.2 Measurement systems

Let us begin by examining the nature of a conventional laboratory measurement system as illustrated by the block diagram shown in fig 1.1. The system consists essentially of three components. The first is the physical or chemical process which is the basis of the measurement, and this can be almost anything from a mouse taking a piece of cheese to carbon-13 atoms absorbing radiofrequency energy at a specified frequency in the magnetic field of an NMR system. The last component is some kind of information presentation or display device which presents a feature of the quantity being measured in a form which can be noted by a human operator. This may be a simple chart recorder trace or a display showing a numerical value. Generally in between there is a second component which acts as a signal converter, converting the signal produced by the first component into the signal required by the third. In most cases the signals are of an electrical nature, and the signal converter is an electronic circuit.

As an example of a simple conventional measurement system consider the basic pH meter illustrated in fig 1.2. In this case the first component of the system is the pH electrode in contact with the solution being tested. The third component is a digital display, allowing readings of pH to be observed directly to two decimal places (eg. 7.04). The signal converter component in this case is an electronic circuit which converts the electrical signal generated by the pH electrode into the whatever signals are required to make the digits 7.04 appear on the display.

Let us now consider a computer based laboratory measurement system, using the simple example illustrated in fig 1.3. In this example the first component may be the same as the first component of

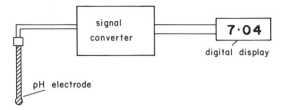

Fig 1.2 The basic components of a typical pH measurement system,
        incorporating an electrode, a signal converter and a display.

a conventional measurement system, thus it might be the pH electrode
as used above. However, instead of just an information display device
we now have a computer receiving the signals generated by the
measurement process. We still require a signal converter system, to
convert the signal produced by the first component into the types of
signals which can be understood by the computer, and this will be a
different kind of electronic circuit from that used in the
conventional measurement system.

    Now using the computer in a most elementary role we could program
the computer (ie. use "software") to produce signals suitable for an
information display, which may be just a display of numbers on a video
screen, a fairly conventional chart record, or even an elaborate
multicoloured diagram annotated with helpful details (such as a
spectrum with peaks highlighted and axes labelled). In that case the
computer based measurement system would provide a similar result to
that obtainable from the conventional measurement system. For example,
if the first component had been the pH electrode, the computer could
be programmed to display the pH on a video screen and we would have an
expensive pH meter.

    However, the computer based system has a number of advantages to
offer. Firstly a computer can be programmed to process data before
displaying a result. Suppose that we use our computerised pH meter for
a system in which a slow chemical reaction is occurring in the sample
at a rate proportional to the concentration of a substance X and that
one of the reaction products causes a change in the pH of the sample
as the reaction proceeds. The computer could be programmed to monitor
the pH variation over a period of time, calculate the rate of the
reaction, and display the concentration of X. This is surely a big
improvement over manually recording a series of pH readings and then
sitting down with a calculator to calculate the quantity of interest.
Secondly a computer may be programmed to store results for later use,
or to compare a result with one obtained earlier. Thus a spectrum
recorded on a computer may be filed away on, say, disk and at a later
time retrieved and overlaid with the spectrum of another sample, to

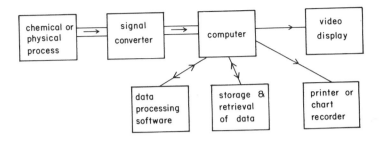

Fig 1.3 The elements of a computer based measurement system, including
both hardware and software components.

aid perhaps in the identification of a sample. Thirdly the computer is
a flexible device, its software may be changed at will. So our pH
meter may be converted into a chloride ion monitor by changing the
electrode to a chloride ion selective electrode and the computer's
program so that it displays the chloride ion concentration. Similarly
a single computer may be connected to several different measurement
devices and with a suitable program may be used to monitor several
different properties at once, such as pH, chloride concentration,
temperature and uv absorbance at 280nm.

Finally a computer based measuring system may also be programmed
to effect and control the conditions in the system on which
measurements are being made. For example, fig 1.4 shows the computer
producing electrical control signals which are then used by a device
(say, a heater) to effect the experimental system in a specified
manner (eg. by changing its temperature), and monitoring electrical
signals produced by a measuring device (say, a thermometer), so that
the computer is able to keep track of the effects of its control
signals.

Each of these aspects of a computer based measuring system has
been dependent on the hardware (ie. the electrical circuits) only to
the extent that signal converters are required to translate the
signals used or produced by various components into those which can be
understood by the computer. But the versatility and power of the the
system really comes from our ability to use software (ie. the
computer's program) to manipulate these signals in an infinitely
extendable variety of ways. This book is largely about the hardware
aspects involved in connecting computers to other components. Although
there will undoubtedly be developments in this field they will
probably fall in the category of technical improvements rather than
revolutionary changes. We discuss software in only a limited way, and
from the much more restricting philosophy that major changes in this
area still lie ahead.

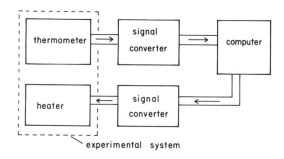

Fig 1.4 A computer based system with control and measurement
       functions.

1.3 Electronic black boxes

   Throughout this book we will be discussing electronic circuits
which form the basis for the hardware connection between a
microcomputer and any external devices. In chapter 2 we examine the
nature of the electrical signals most commonly encountered in the
laboratory and discuss some important aspects of signal transmission.
In chapters 3 & 4 we discuss the types of electronic circuits relevant
to signal converter applications. In all cases it is assumed that the
reader has a basic knowledge of electrical components and electricity,
covering Ohm's law and elementary ac theory, although a detailed
knowledge of electronics is not required. (The symbol R will be used
to indicate resistance values in ohms, k for kilohms and M for
megohms.) Our discussion will be almost entirely confined to
integrated circuits (ICs), which can be regarded as "black box"
electronic devices with precisely defined properties (ie. applying
particular signals at certain connectors of the IC results in
predictable signals appearing at other connectors).

   Most of the ICs mentioned are readily available from a variety of
manufacturers and suppliers, although the complete code numbers used
to identify particular circuits may vary with the manufacturer. For
example, the 709 operational amplifier (see chapter 3) may be
identified by the codes SN72709N, MC1709G, LM709C or a variety of
other codes. Where a device is available from many sources, circuit
diagrams containing the device are labelled using only the device code
(eg. 709) and not with the manufacturer or packaging codes. Where any
ambiguity may result the code of the commercial quality device offered
by National Semiconductor Corporation has been used (generally these
begin with LH, LF or LM depending on whether the device is of hybrid,
BIFET or monolithic construction respectively).

   Many of the devices discussed are available in a variety of
packages. Nearly all of the circuits shown in this book have been used

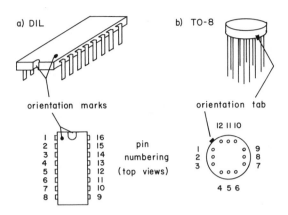

Fig 1.5 Two of the commonly encounter packages used for integrated
circuits. a) the DIL package, and b) the TO-8 can.

in the author's laboratory, where the circuits were constructed using
commercial quality plastic packages (these are usually black) of the
dual-in-line (DIL) configuration illustrated in fig 1.5a. DIL packages
have two rows of connection pins, each pin being 0.1 inch from its
nearest neighbour and the two rows being 0.3 or 0.6 inches apart, and
the number of pins varying from 4 to 40. At no time has the author
found it necessary to use the more expensive ceramic DIL packages
where the plastic packaged devices have been available (some importers
only stock the ceramic packaged devices), nor has the author
deliberately purchased any devices specified to higher than commercial
quality (military and certain other quality devices are specified over
wider operating temperature ranges and may have a tighter spread of
characteristics than their commercial equivalents). A variety of other
packages are available for many analog devices, the 12 pin TO-8 can
shown in fig 1.5b being one of the most commonly encountered. Some of
the devices discussed are only available in this form.

1.4 A practical footnote

    This book was not conceived as a "do-it-yourself" guide to
building computer interace systems, although hopefully some readers
may find that the subject is not as complex as they may have thought
and so be encouraged to do just that. For such readers the following
brief practical comments may be of assistance. The majority of our
circuits have been constructed on pre-drilled IC circuit boards (we
use Eurocards) using "wire wrapping", a technique by which wires are
wrapped without soldering around the pins of IC sockets to convey
signals from one part of a circuit to another. This technique is to be
highly recommended as only a couple of inexpensive tools are required
and wiring errors can be easily rectified. Unless a considerable

amount of test gear and expertise is available the "insulation displacement" form of wire wrapping (in which plastic covered wire is wrapped around the socket pins and the sharp edges of the pins cuts through the plastic to make contact with the wire) is not advised. The time wasted in finding a faulty joint can more than offset the time spent in conventional wire wrapping (where the insulation is removed from the wire before wrapping). For high speed (>10 MHz) and fast TTL (see chapter 4) we have preferred to rely on printed circuit boards.

Many of the systems described involve multiwire connections between a microcomputer and an auxiliary electronic circuit such as an interface system. For such connections the author is a convinced advocate of "insulation displacement" connection systems (available at relatively low cost from most electronic hardware supply houses). In these the connection between the conductor wire and the connector pin is made by teeth (tines) on the connector which are forced through the plastic insulation around the wire, usually with the aid of a simple bench vice. The use of this technique almost forces a high degree of neatness as special soft plastic covered ribbon cables have to be used.

Most microcomputers are powered from the mains, so don't poke around inside with screwdrivers and bits of wire while the computer is still plugged in - the repair costs could be as much as a new computer. Many of the cheaper micros are low voltage devices powered from separate transformer/rectifier units. In either case be careful about taking power from the micro to drive external circuits like interfaces, some manufacturers have provided power supplies which are barely adequate to run the computer let alone any extra circuits. Although the Apple is an exception (it was obviously designed to have extra circuits plugged into its IO slots), with most micros it is usually wise to provide separate power supplies for interfaces using more than half a dozen TTL ICs or equivalent, and in such cases it is particularly important to ensure that the micro does not produce interference (see chapter 2) in any analog interface circuits. TVs and video monitors tend to radiate a fair amount of high frequency signal. The 0 V connections of the micro and the externally powered devices will need to be connected together but some care should be taken about whether the micro can be grounded - certainly check for any voltage between the micro's 0 V level and mains ground first, and if its more than a few millivolts be careful about connecting to ground. If a separate ground connection for the micro has to be provided then this should be done at the power supply circuit and not from a point in the middle of the logic circuits. And always be aware that dabbling inside the computer generally invalidates any warranty it had.

CHAPTER 2

THE BASICS OF LABORATORY SIGNALS

Parameters measured in laboratories cover a wide range of
physical properties and effects, such as luminescence, electromagnetic
energy absorption (uv, ir, nmr, etc.), electrochemical potential,
temperature, pressure, radioactivity, etc., in addition to simple
coordinate changes, such as distance, angle, velocity or acceleration.
However, virtually all such physical characteristics may be converted
into some form of electrical signal by an appropriate energy
conversion device known as a transducer. Some typical transducers are
listed in table 2.1, along with the nature of the particular energy
stimulus for which each is designed.

2.1 Transducers

Transducers fall into one of two primary categories: charge
generating transducers and impedance transducers. The common feature
of charge generating transducers is illustrated in fig 2.1. A circuit

Table 2.1   Examples of laboratory transducers

Transducer	Energy stimulus
Coil	changing magnetic field
Electrochemical electrode	chemical potential
Ionisation chamber	ionising radiation
Microphone	sound/vibration
Photomultiplier	uv/visible photons
Photodiode	light
Piezo-electric crystal	pressure
Resistance thermometer	temperature
Strain gauge	deformation
Thermocouple	temperature, ir

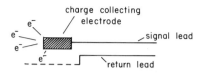

Fig 2.1   The principle of a charge generating transducer, in which an
          electrode collects charge (in this case electrons) liberated
          by some physical process.

element known as an electrode gathers charge liberated by some
physical process and so becomes positively or negatively charged.
Unless we are concerned with electrostatic measurements, a second
pathway is required to return charge to the environment of the
transducer. This is usually achieved through a "ground" or "earth"
connection at a second electrode, although it is possible to "float"
systems at potentials other than ground.

    The effect of this gathered charge is determined in part by the
electrical impedance of any measuring circuit connected to the
electrodes' signal and return leads, and we shall consider this aspect
of the measurement process in section 2.3. For the present we merely
observe that the electrons collected at the electrode may be allowed
to flow along the signal lead, in which case they constitute an
electric current and the transducer provides a source of current.
Alternatively the electrons may be prevented from flowing along the
signal lead - by the presence of a very high impedance to current flow
between the signal and return leads - in which case the potential of
the electrode is changed by the process of collecting the charge. The
voltage change produced on the electrode (or more precisely the
difference in potential of the electrode before and after charge
collection) may be calculated using the equation

$$V = Q/C_T \qquad\qquad (2.1)$$

where Q is the charge collected,
      C_T is the capacity of the transducer electrode, and
      V is the voltage change produced.

Thus a transducer of this class may act as a source of current or
voltage, depending on the technique selected for the measurement of
the charge collected. Examples of this first class of transducer
include devices such as photomultiplier tubes, ionisation chambers and
semiconductor detectors for nuclear particles.

    The second major class of transducer is the impedance transducer,
in which the electrical impedance of the device is influenced either
by its physical deformation, such as the stretching of a resistive

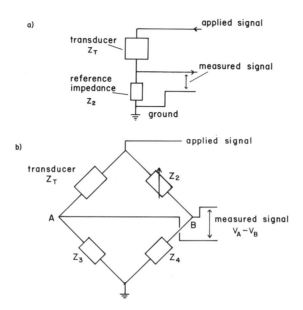

Fig 2.2   Common arrangements for measurements using an impedance
           transducer which undergoes an impedance change as a result of
           a physical stimulus. (a) A simple circuit for measurement of
           impedance relative to a reference impedance. (b) A bridge
           circuit for the measurement of small changes in the
           transducer's impedance.

wire, or by some property of its surroundings, such as the
temperature, pressure, light level or magnetic field. In this case the
impedance of the transducer is monitored using a circuit illustrated
in its simplest form in fig 2.2a. A signal is applied to a series
impedance network made up of the transducer and a second impedance of
fixed or adjustable characteristics known as a reference impedance.
The potential at the junction of the two impedances is measured and
provides information about the variation of the transducer impedance
relative to the reference impedance. As the reference impedance may in
practice be the input impedance of a current measuring circuit, this
class of transducer may also be considered to act as either a source
of current or a voltage source. Typical transducers of this class
include resistance thermometers, strain gauges, magnetometer coils,
light sensitive resistors and diodes, and capacitance devices for the
measurement of distance or dielectric constant.

When an impedance transducer is being used for the measurement of
relatively small changes in a physical characteristic, a bridge
circuit is often used so that the "background" level of the measured
signal may be effectively cancelled out or "nulled". A generalised

bridge circuit is shown in fig 2.2b. The variable impedance, Z_2, may be adjusted so that

$$Z_T/Z_3 = Z_2/Z_4$$

in which case the bridge is balanced, the voltages at points A and B are identical, and the measured voltage difference signal, (V_A-V_B), is zero. Any change in the transducer's impedance will now unbalance the bridge and produce a difference signal related to the impedance change. Bridge circuits are particularly simple to implement when the transducer's impedance is purely resistive, as the impedances Z_2-Z_4 may be resistances and the measured and applied signals may be dc voltage levels. Such arrangements are commonly encountered in measurement systems based on strain gauges.

2.2 Measurement signals

We have seen that transducers act as the sources of current or voltage, and of course to these may be added signal sources for which the property being measured is of an electrical nature (eg. electrochemical cell potentials, conductivities, etc.). However, the signals likely to be met in a laboratory may also have an important time dependence, either because the effect responsible for the transducer output exhibits a time dependence (eg. a vibration or a repeating or decaying signal), or because the signal may be varied in some way to assist in the measurement (eg. the modulation of the magnetic field in magnetic resonance spectroscopy). The types of signals most commonly encountered in laboratory instrumentation are summarised in fig 2.3.

Figure 2.3a illustrates a constant (or slowly varying) voltage or current which has a time average value which is non-zero. This kind of signal is called a direct current, dc, signal, even when one is talking about a voltage or a signal which does vary with time. A dc signal is produced, for example, by a transducer monitoring temperature or pressure in a static system. In this case the quantity of importance is the magnitude of the signal and the signal may be electronically filtered to remove short term variations, such as the effects of noise or interference (see section 2.4). In fig 2.3b a representation of an oscillating or alternating (ac) current or voltage is shown, and the principle feature of this type of signal is that its mean value is zero. Signals of this nature may be found in a radiofrequency receiver or an ac bridge circuit for the measurement of the conductivity of a solution. Probably the most common ac signals are those which have a sinusoidal time dependence, eg.

$$V = V_0 Sin(2\pi ft) \qquad\qquad (2.2)$$

where V is the signal voltage at time t,
      V_0 is the peak amplitude and

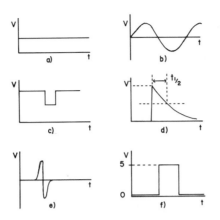

Fig 2.3 Types of signals commonly encountered in laboratory
         instrumentation. In each case the variable signal may be a
         current or a voltage. a) a dc signal, b) an ac signal, c) a
         unipolar pulse, d) a decaying signal or a long tail pulse, e)
         a bipolar pulse, and f) a logic pulse.

         f is the normal frequency of the oscillation
            in hertz.

In this case either the absolute magnitude or the frequency (or both)
of the signal may be the important parameter for measurement, and
filtration of the signal to remove variations of a different frequency
may be accomplished easily only if the frequency range over which the
signal may vary is limited. The magnitude of an ac signal may be
expressed in terms of its peak values (V_0) or its root mean square
(RMS) value, which for a sinusoidal signal is about $0.707V_0$.

     Figure 2.3c illustrates a pulse signal. A current pulse of this
kind is produced by a photomultiplier tube detecting brief flashes of
light or single photons. In this case filtration of the signal is
virtually impossible in fast systems and alternative techniques have
to be used to minimise the effects of noise and interference. The
important parameter associated with pulse signals may be the magnitude
of the pulse (which may be its height or its integrated area) or the
number of pulses per second. It is the difficulty in measuring the
height of very narrow pulses which often leads to the conversion of
such pulses into the "long tail pulse" form of signal illustrated in
fig 2.3d. Generally the tail of the signal has a characteristic
mathematical form, such as an exponential decay. The conversion may be
carried out by the integration of fast pulses, as discussed in section
3.9, although the same form of signal may also be characteristic of
measured physical properties - such as the intensity of a laser flash,

or the number of bacteria surviving after a chemical has been added to their growth medium.

The long tail signal is produced as a voltage pulse from a charge sensitive transducer (see eqn (2.1)) when the charge is collected rapidly, but bleeds away from the collecting electrode more slowly (usually through a high resistance). Examples include photomultipliers detecting scintillations and X-ray and nuclear particle detectors. For this signal the important parameter may be the pulse height or the time of decay of the signal, which is usually taken to be the time interval between the time at the peak of the signal and the time at which the signal falls to one half (or sometimes 1/e) of its peak value. While it can be illustrated in a single diagram this type of signal actually covers a wider range of measurement techniques than most of the others, simply because of the very wide range of time scales which may be involved. For example, measurements can be made on the half-life of a radioisotope, which may be minutes or hours, or on the lifetime of a fluorescent species which may be a few nanoseconds. As a result this kind of signal may be the most difficult to measure accurately, particularly as its measurement relies upon maintaining the fidelity of the transducer output through many stages of electronic signal processing, which is difficult to achieve over a wide frequency range.

The pulse signal shown in fig 2.3c is actually a unipolar pulse, as the signal value is always on only one side of the background voltage or current (which may, of course, be zero). This kind of pulse shares a number of characteristics with a dc signal, in spite of the fact that it may involve very fast changes of voltage or current. An alternative type of pulse is the bipolar pulse, shown in fig 2.3e, which has a time-averaged signal level of zero and so may be regarded as a kind of ac signal. Bipolar pulses, in common with other ac signals, may be passed through capacitors which block dc signals and, as we shall see, this "ac coupling" of signals can offer certain advantages. Unipolar pulses have to be treated more carefully: attempting to pass them through a capacitor, for example, can result in their conversion into bipolar pulses and hence a change in the time-averaged signal level.

All of the above signals are examples of analog signals, because their magnitude carries information which can represent the value of a continuously variable quantity. However, an increasingly common type of signal associated with laboratory instrumentation is the logic level signal. An example is provided by the logic pulse shown in fig 2.3f. A logic pulse consists of a transition from one specified logic level to another and back again. In most cases the signal levels involved are 0 V and +5 V, although 0 V and +10 V are also common. (Logic level signals will be dealt with in more detail in chapter 4.) In logic pulse systems the important parameter may be the presence or absence of a pulse, the number of pulses per second or the width of

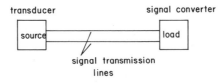

Fig 2.4 The connection of a transducer to a signal converter. In many
         cases one of the transmission lines may be grounded.

each individual pulse (and the latter may be used in an analog sense).
While the number of transducers which generate logic pulses directly
is limited (eg. to mechanical devices and some trigger circuits), many
pulse producing transducers can be fitted with microcircuit converters
which convert current or voltage pulses above a certain threshold into
logic pulses (as will be illustrated in fig 4.16). Photon counting
photomultiplier systems and chanelltron multipliers are ideal
candidates for this treatment because the logic pulse signals are
virtually immune from noise or interference and may be transmitted
over long distances without loss.

2.3 The transducer connection

     The electrical signal produced by a transducer is of value only
if the signal characteristic of interest can be measured. To make any
kind of electrical measurement requires the connection of the
transducer's output to the input of a signal converter, a device which
converts the transducer's electrical signal into a signal which may be
more easily presented to and understood by man. The signal converter's
output may drive a display device which may take any convenient form,
from a moving needle meter to a picture (provided, say, by an
oscilloscope trace) or a sophisticated computer display. The
transducer connection is illustrated schematically in fig 2.4, where
two wires, known as signal transmission lines, carry the transducer
output to the signal converter input, which acts as a load for the
signal's energy. It is the signal on these transimssion lines which
forms the basis of all subsequent measurements, so that, in spite of
the fact that these lines are all too frequently neglected in
laboratory work, we will consider them in some detail. Much of the
discussion also applies to signal transmission lines connecting one
stage of a converter system to the next.

     For most laboratory signals the transmission system may be
regarded as consisting of the elements shown in fig 2.5. The origin of
the signal is a two terminal signal source with a source series
impedance which is an inherent feature of the source design. The
source may also be associated with a parallel impedance (usually
capacitive), although this is generally of far less importance unless
very high signal frequencies are used. Where the signal source is a

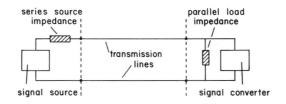

Fig 2.5   The principal features of the general circuit formed by the
          connection of a signal source (such as a transducer) to a
          load (such as a signal converter).

transducer, the source series impedance is generally outside the
control of the user, and may range from a low value (say, 10 R for a
hot filament or 100 R for an ir thermocouple) to a very high value
(say, 10^9 R for a pH electrode or 10^8 R for an illuminated
photomultiplier tube). The signal source and its associated series
impedance form the termination at one end of the transmission line. At
the other end is the load formed by the input circuit of the signal
converter, which may be regarded as a measuring device of infinite
impedance in parallel with a parallel load impedance. As the signal
pathway is from the signal source to the signal converter, the series
source impedance is often referred to as the output impedance of the
source, and the parallel input impedance of the converter as simply
the input impedance of the converter. The user does have control of
the converter's input impedance to the extent that he can select the
most suitable converter input circuit for the system in question.

     An important feature of a transmission line terminated at each
end by devices with impedance is the efficiency with which power is
transferred from one end of the line to the other. We first consider
the amount of power generated by a simple source which is producing a
constant dc voltage, as illustrated using the equivalent circuit shown
schematically in fig 2.6. (This enables the discussion to be carried
out in terms of resistances, rather than complex impedances.) If the
signal source is generating a dc voltage, V_S, and its series
resistance is R_S, then the maximum power generated by the source will
be produced when the maximum current is drawn from it, and this will
occur when the output leads are shorted together. Designating this
short circuit current I_S, then the maximum power generation is

$$P_{max} = V_S * I_S = I_S^2 * R_S = V_S^2 / R_S \qquad (2.3)$$

If the output is now connected to a load resistance, R_L (which may
represent the input resistance of a converter circuit), as illustrated
in fig 2.7, and if the current flow is now I, then the power
dissipated in the source resistance becomes

$$P_S = I^2 * R_S$$

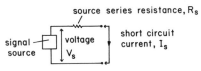

Fig 2.6   Equivalent circuit for a voltage signal source with its
          series source resistance, shown with the output terminals
          shorted so that the maximum current may flow.

and that dissipated in the load resistance R_L becomes

$$P_L = I^2 * R_L$$

The total power being dissipated is now

$$P_T = P_S + P_L$$

$$= I^2 * (R_L + R_S) = V_S^2 / (R_L + R_S)$$

It follows that the power transferred to the resistance R_L is

$$P_L = (V_S^2 * R_L) / (R_L + R_S)^2 \tag{2.4}$$

which, when combined with eqn (2.3), gives

$$P_L = P_{max} R_L R_S / (R_L + R_S)^2 \tag{2.5}$$

If $R_L \gg R_S$ then the power transferred to R_L approximates to

$$P_L = P_{max} * R_S / R_L$$

Thus if R_S = 50 R and R_L = 1 M then $P_L = 5 * 10^{-4} * P_{max}$. Similarly if
$R_L \ll R_S$ then

$$P_L = P_{max} * R_L / R_S$$

and if R_S = 1 M and R_L = 50 R then $P_L = 5 * 10^{-4} * P_{max}$.

To find the maximum power transferable to R_L for a fixed value of
R_S it is necessary to find the value of P_L when the partial
derivative, dP_L/dR_L, is zero. Differentiating eqn (2.5) we find

$$dP_L/dR_L = P_{max} R_S (R_S - R_L) / (R_L + R_S)^3 \tag{2.6}$$

Equating this to zero we find that the maximum transfer of power
occurs when $R_L = R_S$ and under these conditions

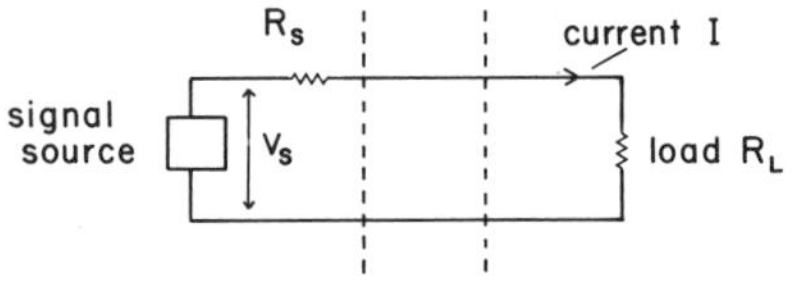

Fig 2.7   Equivalent circuit for the source of fig 2.6 connected to a
          load resistance R_L.

$$P_L = P_{max}/4$$

While it may seem somewhat disappointing that the best we can hope for
is the transfer of 25% of the maximum available power from the signal
source to the load, R_L, it is perhaps somewhat reassuring to realise
that this is 50% of the power actually being generated by the signal
source under the conditions of the circuit in fig 2.7.

     A similar argument may be applied to the circuit of fig 2.5, in
which the resistance R_S is replaced by the source impedance and the
resistance R_L by the input impedance of the signal converter. In the
general case of complex impedances it is found that, for maximum power
transfer of an ac signal, the source impedance must equal the complex
conjugate of the load impedance. This leads to one of the fundamental
rules of electrical signal transmission – that for the maximum
transfer of power between a signal source and a load, the impedances
of the source and the load must be matched, where "matching" implies
equating resistances or the conjugates of impedances, depending on the
nature of the signal. This rule is of particular importance in
connection with the measurement of transducer output signals because
the limit of sensitivity of the transducer is largely determined by
the smallest amount of energy which can be measured at the input of
the signal converter circuit. Clearly this will represent a higher
proportion of the transducer's capability if the impedances are
matched than if less than 0.005% of the transducer's output is
actually usable at the signal converter's input. Some examples of
common transducers are given in table 2.2, along with their
characteristic output impedances and typical signal levels.

     In spite of this demonstration that the maximum power transfer
occurs when the impedances at the ends of a transmission line are
matched, it must be understood that it is not always necessary or
desirable to operate under conditions of maximum power transfer. For
example, if we wish to measure a dc voltage generated by a source,
then using a measurement device of equal impedance will result in
current being drawn through the source series resistance, with the
consequent potential drop across that resistance introducing an error
into the voltage measurement. In fact under these conditions the
voltage at the input terminals of the measurement device will be half

Table 2.2   Examples of common transducers for
            laboratory measurement purposes

transducer	output impedance	typical signal range
hot filament	10^1 R	0.1 - 10 V
ion chamber	10^7 R	0.1 - 100 nA
ir thermocouple	10^2 R	1 - 100 nV
pellistor	10^2 R	1 - 100 mV
pH electrode	10^9 R	-1 - +1 V
photodiode	10^2 R	1 nA - 1 mA
photomultiplier	10^8 R	1 - 1000 nA
piezo crystal	10^4 R	1 - 100 mV
pyroelectric fet	10^3 R	1 - 1000 mV
strain gauge	10^4 R	0.1 - 10 mV
thermistor	10^4 R	1 - 100 mV

of that being generated by the source. To measure 100% of the source
voltage we would need a measuring device of infinite input impedance,
and in this case no power would be transferred from the source to the
measuring device. In general it is desirable to measure dc voltages
with a measuring device having a much larger input impedance than the
series impedance of the source - this allows the measurment to be made
with the highest accuracy and without the necessity of correcting for
voltage drops across the sources series impedance. BUT, because the
energy transfer involved is very small, there is the possibility of
interference by other small energies which may find their way into the
measuring device (see section 2.4).

On the other hand dc current measurement would be better
accomplished using a measuring device having a very low input
impedance compared with the source's series impedance - for under
these conditions the current flow is maximised. Again little power is
transferred to the measuring device because, although current flows
through it, the potential drop which results is small and little power
is dissipated in the measuring device - most of the power being
dissipated in the source series resistance.

While ac signals at low frequencies can generally be treated in
much the same way as dc signals, at high frequencies (>1 MHz) there is
not quite as much choice. If impedances at the ends of a transmission
line are not matched, then high frequency signals may be reflected
from the terminations and their energy dissipated in the conductors of
the transmission line. (This can cause high voltages to be developed -
higher than those being applied by the source - and these can be

dangerous if breakdown of the cabling insulation occurs.) At
frequencies above 100 MHz virtually every signal handling conductor
needs to be treated as a transmission line and to have its end
impedances matched.

2.4 Noise and interference

Two factors which are of importance in determining the limit of
sensitivity of a transducer system are the often confused phenomena o
noise and interference. Noise consists of fluctuations in the signal
level and may be a feature inherent in the nature of the signal or
superimposed on the signal by circuit elements in the source and
measurement circuits. In properly designed systems the noise
introduced into the system by the measurement electronics should be
small compared with that originating within the transducer and its
associated series resistance, so that noise may normally be regarded
as a property of the signal source. Interference on the other hand
consists of fluctuations of the signal level caused by the addition o
spurious signals from sources not directly involved with the
measurement process. One of the commonest forms of interference is the
induction of small currents in the signal transmission lines by the
interaction of electric and magnetic fields with the conductors which
make up the transmission lines.

The measurement system and sources of noise and interference are
illustrated in fig 2.8. We will briefly consider each of these proble
areas in turn, although it should be appreciated that our purpose is
to introduce the nature of the problem rather than to consider the
subject in great detail. Firstly there is noise associated with the
nature of the signal itself. Actually there are two kinds of noise
which should come under this heading, but one of these - fluctuation
of the physical property being probed by the transducer - is more
conveniently described as background signal variation. This will not
be further considered in the present context, as proper (although
sometimes very time consuming) design of the physical system involved
can reduce this source of noise to insignificant levels. The more
limiting source of signal noise is "shot noise", which is the inheren
fluctuation of the signal level due to the statistical nature of the
electric current (ie. charge movement) produced within the
transducer.

The nature of shot noise may be most readily understood by
considering a current, I, measured by detecting the number of
electrons, x, arriving at the signal converter input in time t. As th
rate of arrival is x/t it follows that the current should be xe/t,
where e is the electronic charge. However, the statistical variation
in the number of electrons arriving in time t is given by the varianc
in x, which we can take as $x^{1/2}$, so that the number arriving is more
properly written as:

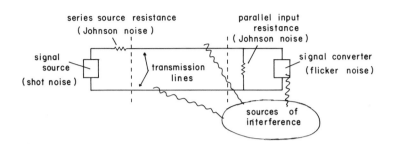

Fig 2.8   Equivalent circuit for a typical transducer measurement
system, illustrating the major contributors to noise and
interference signals:- the signal source, the source series
impedance, the transmission lines, and the signal converter
circuit.

$$x + x^{1/2}$$

and the current as

$$(x + x^{1/2})e/t$$

The noise (ie. the statistical variation) in the measured current may
be quantified by its variance, so that as a fraction of the mean
current the noise is

$$(x^{1/2}e/t)/(xe/t) = x^{1/2}/x = x^{-1/2}$$

If the number of electrons arriving in time t is now increased to 4x,
the current may now be written

$$(4x + 2x^{1/2})e/t$$

and the fractional noise in the current is reduced to

$$2x^{1/2}/4x = x^{-1/2}/2$$

In each case the noise is an example of shot noise, and the
calculations illustrate that the shot noise is reduced by increasing
the number of physical events contributing to the measurement - in
this case electrons recorded by the signal converter.

It is instructive to consider one further example in which the
mean current remains xe/t, but the time interval over which the
arrival of electrons at the signal converter input is recorded is
increased from t to 4t. In this case the number of electrons to arrive
may be written as $4x+2x^{1/2}$, and the current becomes

$$(4x + 2x^{1/2})e/4t$$

The fractional noise in the current is now

$$2x^{1/2}/4x = x^{-1/2}/2$$

illustrating the important rule that it is the number of physical events recorded which governs the shot noise, rather than the rate at which the events occur. Often shot noise effects may be reduced by making measurements over a longer period of time.

Shot noise effects are of particular importance in event counting systems, such as single photon spectrometers or single ion mass spectrometers, but other sources of statistical fluctuation in signal level are to be found in more humble systems. For example, the gain of a typical photomultiplier tube may be quoted as $3.0*10^6$, but this represents an average gain for electrons ejected from the photocathode. Many photoelectrons will be multiplied by smaller values, and many others by larger values. Thus the anode current of a photomultiplier exposed to a steady light level will show variations in magnitude due to shot noise, and these variations will represent a larger fraction of the mean signal level, and so appear as a higher noise level, as the light level is reduced.

Shot noise is a characteristic which arises because of the discrete nature of electric current (or photons), and so its effects are discernible for all devices which utilise currents. The magnitude of the shot noise inherent in the flow of a current is given by

$$\langle I_n \rangle = (2eIB)^{1/2} \text{ A} \qquad\qquad (2.7)$$

where e is the electronic charge
    I is the average dc current, and
    B is the frequency interval (bandwidth) over which   the shot noise current is required.

With B in hertz, eqn (2.7) becomes

$$\langle I_n \rangle = 1.789*10^{-10}(IB)^{1/2} \text{ A}$$

The noise level is a function of the frequency interval rather than the frequency, and for a given frequency interval is independent of the frequency. Consequently shot noise is known as white noise, being present at the same intensity at all frequencies. However, because bandwidths are generally a larger number of hertz for circuits handling high signal frequencies than for those handling low frequencies, shot noise produces its greatest practical effect in high frequency systems. For this reason high frequency analog signal handling systems often employ bandwidth narrowing techniques to minimise noise levels.

A second important source of noise is the thermal excitation of electrons in resistive elements of the ciruits associated with signal handling. This is called Johnson noise and appears as a fluctuating voltage across a resistive component — whether or not that component is passing a current. Its magnitude may be expressed as the mean value of the noise voltage, $\langle V_n \rangle$, within a specified frequency range, using

$$\langle V_n \rangle = (4kTRB)^{1/2} \text{ V} \qquad (2.8)$$

where R is the resistance of the element
  k is the Boltzmann constant
  T is the absolute temperature of the element, and
  B is the frequency interval over which the noise voltage is
  required.

With R in ohms and B in hertz, eqn (2.8) becomes

$$\langle V_n \rangle = 1.3 * 10^{-10} (RB)^{1/2} \text{ V at 300 K}$$

As was the case with shot noise, the Johnson noise level is a function of the frequency interval rather than the frequency, so it is a white noise and contributes the same noise level in each equal frequency interval, again producing the greatest practical effect in larger bandwidth, high frequency circuits. It is clear from eqn (2.8) that its effects are minimised by the choice of low value resistances for circuit elements handling very small signals (such as the signal converter input circuit), and by restricting the range of frequencies accepted by the signal converter. Unfortunately it is not always possible to implement these desirable features into a practical converter circuit, and compromise is sometimes required.

Shot noise and Johnson noise are factors which are inherent in the physical nature of current and the resistance to current flow respectively. However, there is a third source of noise which can be distinguished and which arises from the circuit elements used to handle electrical signals (or indeed any other kind of signals). This type of noise is called flicker noise or 1/f noise, and has an inverse frequency dependence — being most troublesome for low frequency or dc handling systems. Flicker noise appears to originate from imperfections in manufactured circuit elements, such as resistors and transistors, and its magnitude increases with the current flowing through the device (unlike shot and Johnson noise).

Flicker noise is normally considered on a purely empirical basis and for modern cicruit elements, such as transistors and integrated circuit amplifiers, is lumped together with shot noise arising from the small currents used in the bases or gates of component transistors. This form of noise may be regarded as consisting of a noise voltage and a noise current present at the input of the amplifier, and so subject to the same degree of amplification as the

signal being processed, (the noise is said to be "referred to input").
Circuits accepting signals from low impedance sources (<1 k) tend to
have their output noise level dominated by the voltage noise at the
input, while those with high impedance inputs (>1 M) have their output
noise level determined principally by the input noise current flowing
through the high input impedance.

The noise levels in standard IC amplifiers can be surprisingly
high, the values for the famous 741 op-amp (see chapter 3) being about
70 nV $Hz^{-1/2}$ and 1 pA $Hz^{-1/2}$ (at a frequency of 10 Hz). Manufacturers
of "low noise" IC amplifiers publish data sheets containing details of
the noise levels associated with their devices, and for typical high
quality amplifiers figures of the order of 10 nV $Hz^{-1/2}$ and 0.1 pA
$Hz^{-1/2}$ (at a frequency above 10 Hz) are to be expected, although these
figures may rise at lower frequencies. This corresponds to a total
noise level of more than a microvolt voltage noise and more than 100
pA current noise when integrated over the frequency range 0.1 - 10 Hz.
It must be remembered that these figures are referred to the input and
so apply before amplification. Newer devices may be expected to
improve on these figure by about an order of magnitude, but
nevertheless such noise levels still present a fundamental limitation
to the precision with which small, low frequency signals may be
measured.

The three important forms of noise outlined above are essentially
independent of one another and the total noise level of a system can
be estimated from the square root of the sum of the squares of the
individual contributions. Generally only one of the noise terms will
dominate the total noise, but an example showing all four terms is
illustrated in fig 2.9. A current source is passing its current, I_S,
into a resistor, R_S, producing a voltage drop which is to be amplified
by a voltage amplifier. The noise levels can be considered
independently of any gain we ascribe to the amplifier. There are four
noise sources:

1). Shot noise, $\langle I_N \rangle$, in the signal current, I_S.
2) Johnson noise, $\langle V_N \rangle$, in the resistor R_S.
3) Amplifier voltage noise, $\langle v_n \rangle$, (flicker dominated at low
   frequencies), and
4) amplifier current noise, $\langle i_n \rangle$, (also flicker dominated at low
   frequencies).

The two current noise terms produce voltage noises by passing
through R_S, and the total noise may be written:

$$\langle V_T \rangle^2 = (\langle I_N \rangle R_S)^2 + \langle V_N \rangle^2 + \langle v_n \rangle^2 + (\langle i_n \rangle R_S)^2$$

While amplifier noise terms can generally be derived from
manufacturers data, some care is needed in estimating the correct
values of the shot and Johnson noise terms. For example, the shot

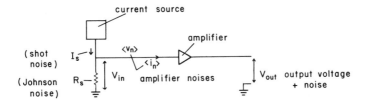

Fig 2.9 A hypothetical system illustrating four sources of noise: shot
         noise in the source current; Johnson noise in the resistor R_s;
         and the amplifier voltage and current noises.

noise in a photomultiplier anode current is very much greater than
that given by eqn (2.7), because the anode current is a highly
amplified version of the photocathode current and contains at least
the same fractional noise as that present in the photocathode current.
Thus for a tube operating with a gain of 10^6 and an anode current of
10^{-6} A the shot noise in the anode current for a bandwidth of 1 Hz is
actually $6*10^{-10}$ A, or 0.06%, rather than the $6*10^{-13}$ A, or 0.6 ppm,
suggested by the application of eqn (2.7) to the anode current alone.
So the calculation of shot noise for currents arising from transducers
with built-in gain must be carried out using the unamplified current
value. Furthermore although photomultiplier tubes present a very high
resistance pathway to current flow, they do so by virtue of a vacuum
between their electrodes rather than a resistive material, so that
noise arising from thermal excitation of electrons in resistive
materials does not contribute to photomultiplier noise and these
devices have no Johnson noise component. (Unfortunately they make up
for this by thermal emission from the photocathode giving rise to a
dark current).

2.5 Minimising interference

     Interference is most frequently caused by small currents induced
in the signal transmission lines as a result of the interactions
between changing electromagnetic fields and the conductors. If the
mean value of this induced current is written $\langle i_i \rangle$, then when this
current flows through, say, the input resistance, R_{in}, of the signal
converter circuit it results in a voltage variation given by

$$\langle V_i \rangle = \langle i_i \rangle R_{in}$$

It follows that interference is most likely to be serious when R_{in} is
large. For example, with $\langle i_i \rangle = 10^{-8}$ A, an input resistance of 1 M gives
rise to $\langle V_i \rangle = 10$ mV, while an input resistance of 50 R gives rise to
$\langle V_i \rangle = 0.0005$ mV. The importance of the values in relation to the
measurement of a 10 mV signal is self-evident.

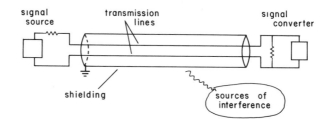

Fig 2.10 The use of shielding to minimise the induction of
         interference currents in transmission lines.

When the source and the signal converter have low impedances
interference is rarely a problem, the precision with which a signal
may be measured being limited by noise in the signal and flicker noise
introduced by the circuit elements of the converter. Unfortunately
many widely used transducers have a very high output impedance and so
require signal converters with a high input impedance. To minimise the
effects of interference in such systems the first priority should be
the elimination of any local sources of interference, as this often
significantly relieves the problem. Thus any troublesome local
thermostats and motors should be fitted with interference suppressors
However, there are other steps which can be taken to minimise the
effects of interference generated outside local control, and most of
these steps concentrate on the points at which the interference
currents are introduced to the transmission lines and the signal
converter input.

The first approach is to minimise the magnitudes of currents
induced on the signal transmission lines by keeping these as short as
possible, preferably less than a few centimetres long. Secondly, the
transmission lines should be shielded (or screened) wherever possible
to prevent electromagnetic fields reaching the conductors. While
conventional coaxial cable, consisting of an inner conductor and an
outer shielding braid connected to ground, is better than nothing, a
preferred technique is to shield both conductors of the transmission
line - even if one of these is grounded. (Interference currents are
generally unaware of the purpose served by the particular piece of
conducting material they inhabit.) Shielding should be connected as
illustrated in fig 2.10 and grounded at one end only, to prevent
induced currents flowing along the shielding and inducing currents in
the adjacent transmission lines.

In some cases long lengths of transmission lines are unavoidable
and shielded conductors (which are never perfectly shielded) are
inadequate. In such cases the easiest solution is often to move a part
of the signal converter input circuit from the converter end of the
line to the signal source end. This can be achieved by including an

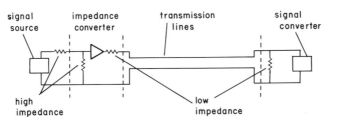

Fig 2.11 The use of an impedance converter positioned close to a high
impedance signal source to provide a low impedance output for
connection to long transmission lines.

impedance converter circuit (see chapter 3) as close as possible to
the signal source, and converting the high impedance output of the
source to a low impedance output of the impedance converter, as
illustrated in fig 2.11. If the signal converter input is now a low
impedance input, interference pickup on the transmission line is
likely to be small even on many metres of transmission line.

All that is needed to achieve successful interference pick up is
a conducting pathway from the laboratory environment to the input
circuitry of the signal converter. Most sensitive signal converter
systems are likely to be housed in metal cases, although conducting
paints for plastic cases and cases constructed of plastics laminated
with a conducting layer are also available. In sensibly designed
systems, these cases can provide their own shielding for the wiring
inside, but it should not be forgotten that in many systems conductors
other than the transmission lines pass through the case walls. In
particular a mains (or line) supply may be connected to miles of
unshielded conductor acting as a magnificent antenna for all the
interferring signals in town. Wrapping a 1 metre length of plastic
covered wire around the mains lead of an oscilloscope and viewing the
signal picked up on the wire will demonstrate the magnitude of the
mains borne interference problem in any laboratory. Usually there is
no shortage of interference in the 1-100 MHz region. A wide selection
of rf line filters is available at relatively low cost and many of
these are reasonably effective at preventing interferring signals
being radiated from mains conductors within the instrument case.
Unfortunately ground wires can still be a serious source of
interference, and it may be helpful to ensure that only one connection
is made between the ground points of individual circuits and the
external ground provided by the mains ground wire.

Low level dc supply lines and chart recorder output wires are
other, sometimes neglected, sources of interference pickup, and the
solution is often to be found in screening the cables and installing
ferrite hoop filters on the wires inside the case and as close to the

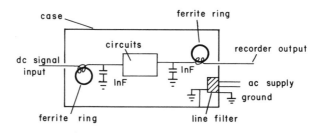

Fig 2.12 Typical techniques for minimising the introduction of
         interference signals to instrumental circuits by suppressing
         leads entering the instrument case. (Both recorder lines
         should receive the same treatment, as should both input lines
         - even where one is nominally a ground line.)

entry point as possible, followed immediately by a small value
capacitor, say 1 nF, as illustrated in fig 2.12. (Incidentally, using
a ferrite hoop without a following capacitor can make matters worse
rather than better, as the hoop may become an effective radiator of
interference.) It should also be borne in mind that the multiphase
wiring of modern buildings, fluorescent lighting and the high currents
used in much modern laboratory equipment result in a number of perhaps
unexpected phenomena. For example, fluorescent lighting produces a
significant amount of radiation at double the mains frequency.
Furthermore, it is not uncommon to find dc level differences of
several hundred mV between the ground pins of two mains outlets in the
same laboratory, and to find different patterns of rf coming from each
outlet. It is usually easier to tackle mains and ground borne
interference if the mains connections of each unit of a system can be
made to a single outlet.

2.6 Signal-to-noise ratio

     A parameter which is most useful in defining the quality of a
complete measuring system is the signal-to-noise ratio of a specified
signal. For example, the limit of measurement for many systems is
defined as the signal level which can be measured with a signal-to-
noise ratio of 2:1 (or just 2). The signal-to-noise ratio of a system
is normally determined at the output of the signal converter; however,
it is convenient to introduce the quantity formally at this point now
that we have given some thought to the origins of noise and
interference. In the present context the word noise is being used to
mean the fluctuations in the output of the measuring system, deriving
from all sources, and is quantified as the variance of the output
signal. The signal-to-noise ratio (SNR) may be defined as the ratio of
the mean output of the measuring system to the variance of that
output, ie

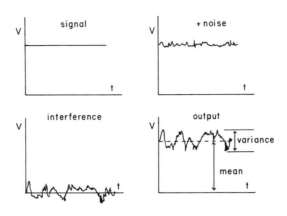

Fig 2.13 The effect of adding to a constant level signal, (a),
        contributions of noise, (b), and interference, (c), to
        produce the kind of signal, (d), usually encountered in
        laboratory measurements.

$$SNR = \langle V_S \rangle / \langle V_N \rangle$$

where $\langle V_S \rangle$ and $\langle V_N \rangle$ are the mean values of the signal voltage and the
noise voltage respectively. An analogous expression may be written for
the SNR of current output systems, although these are less common. The
output in question may be a nominally constant signal level (such as
the value recorded by a pH meter with the probe in a buffer solution),
or a difference between values of a signal level which changes while
the measurement is being made (the height of a chromatographic peak
for example). The contributions of signal, noise and interference to
the output of a dc measuring system are illustrated schematically in
fig 2.13.

Signal-to-noise ratios are often expressed using the dB scale, in
which case

$$SNR = 10 \, \log_{10}(\langle V_S \rangle^2 / \langle V_N \rangle^2) \text{ dB}$$

but whichever system of units is used it is important to specify the
bandwidth over which the SNR applies. For the measurement of a signal
of closely defined frequency, the use of a measuring system with a
wide frequency response allows an increase in the observed value of
$\langle V_N \rangle$ without any improvement in $\langle V_S \rangle$, thus producing a degraded SNR.
The situation is particularly critical for the measurement of dc
signals because of the contribution of $1/f$ noise. While it may be
tempting to assume that f has no relevance to dc measurements, think
again. One generally requires laboratory measurements to be completed
within a finite time, and even if we allow 100 s for a particular

measurement, this corresponds to an f of 0.01 Hz — so a bandwidth of 1 Hz for the measuring device is 100 times greater than is needed and allows $\langle V_N \rangle$ to be greater than necessary.

The optimistic reader may feel that these problems of noise and interference have been somewhat laboured. However, there is a good reason for this. In later chapters we will be considering the recording of signal levels by computer. A computer can collect the value of a signal at a particular microsecond, while a pen recorder records a time-averaged signal (typical pen recorder response time is of the order of 0.5 s) so that any rapid fluctuations in a signal level are usually not apparent on a chart record. A signal which records on a chart recorder as a noiseless line may, when read twice per second by computer, appear as a most disappointing collection of apparently randomn numbers.

2.7 Control signals

Most of the signals we have encountered from measuring devices are also used for control functions or for the production of physical effects required in a laboratory. Examples of some common uses of signals for these purposes are given in table 2.3. The principle differences between the characteristics of signals associated with measurement and control are the magnitudes of the signals and the impedances of the devices which generate or utilise them. Thus the

Table 2.3   Examples of controlled devices

device	operating signal	output
discharge lamp	100 V, 5 A, dc	uv light
filament	10 V, 1 A, dc	red light
heater	mains 5 A	heat
LED[a]	5 V, 5 mA, dc	low-level light
loudspeaker	10 V, 1 A, ac	sound
microwave source	10 V, 100 mA, dc	X-band microwaves[b]
motor	24 V, 2 A, dc	motion
pump	mains 1 A	fluid pressure
stepping motor	5 V, 1 A, pulses	position
ultrasonic source	12 V, 25 mA, ac	ultrasound
x-ray generator	8 kV, 100 mA, dc	X-rays

[a] light emitting diode.   [b] low power modules.

photomultiplier detector of a spectrometer may produce an output of a
few microamperes when a strong signal is being recorded, whereas the
light source of the same spectrometer may require a current of several
amperes, supplied with a potential of many tens of volts.

Signals at the higher levels common for control and power
functions of instrumentation are not normally subject to interference,
although often responsible for causing interference in other systems.
However, noise (ie. fluctuations of the signal level) can present
serious problems, particularly if the fluctuations are likely to
effect some sensitive measurement transducer. Dc signals at low
voltages (<100 V) are relatively easy to stabilise and filter using
voltage regulator ICs and simple RC filters, but at the high voltages
generally required as the bias supplies for photomultiplier and
channeltron detectors, the removal of noise is much more difficult.
Most commercial power supplies in this category are rated for their
noise and ripple (ie. residual variation at the ac frequency used to
generate the high voltage) levels, and these levels are likely to
increase as more current is drawn from the supply. Looking at the
noise level on high voltage lines can be difficult as most
oscilloscopes have maximum input voltages of 500 V or less (whereas
photomultipliers generally require 1000 - 2000 V bias), and it is
usually necessary to divide the examined voltage down using a high
voltage dividing probe.

Ac power signals derived from the mains supply are most likely to
be used for the operation of heaters, lights and simple motors,
whereas ac signals at higher frequencies are used for more exacting
requirements, such as precision conductivity measurements, magnetic
resonance systems, quadrupole mass spectrometric analysis and rf
heating. Non-sinusoidal oscillating signals are also popular; for
example, triangular waveforms are useful in providing short duration,
constant acceleration motions, and square waves are the ideal
repetitive on/off signals for lamps and mechanical devices. On/off
signals with a variable on/off ratio are useful for driving dc motors
at varying speeds without the loss of torque which accompanies voltage
reduction, and motor controllers are available for this purpose.

Ac signals up to mains voltage can be turned on and off using
solid state relays or triacs, and opto-isolated versions of both
devices are available to preserved high reliability insulation between
the switching circuit and the switched signal. Triacs can also be used
for controlling power in ac circuits, although generally only if the
load is resistive (eg. a lamp or heater). Dc signals may be turned on
and off using power transistors or regulator ICs, and the development
of VMOS and DMOS power fets enables this technique to be used for
signal levels beyond 1000 V. Controlling dc levels is most readily
accomplished using variable level regulator ICs, although for voltages
greater than about 100 V it is necessary to use a power fet and to
control its gate voltage. Most of the necessary devices can be found

in the catalogues of electronic component suppliers, and many suppliers will also provide the relevant data sheets on request.

CHAPTER 3

THE ELEMENTS OF ANALOG SIGNAL HANDLING

Most transducers and laboratory signal sources yield output
signals the magnitude of which carries at least a part of the
information required in the measurement process. Such signals are
called analog signals. It follows that most signal converter systems
are required to handle analog signals, at least in the input stages of
the converter circuit. One of the most widely used circuit elements
for handling analog signals is the amplifier, which, as its name
implies, is intended to convert a small signal into a larger one,
often to overcome the inconveniently small magnitude of the signal
generated by a transducer. Early amplifiers were constructed using
thermionic valves, which were physically large and consumed large
amounts of power - most of which was wasted as heat. Later amplifiers
were constructed using transistors, and many special purpose
amplifiers (such as VHF, Very High Frequency, amplifiers) still need
to be assembled from a number of these "discrete" devices. However, in
the majority of cases amplifiers can be more easily constructed using
integrated circuit elements called operational amplifiers (op-amps for
short). Op-amps have in fact become the basic ingredient in nearly all
aspects of analog signal handling and we shall concentrate our
discussion on these devices.

3.1 Op-amps

Operational amplifiers are supplied in a variety of packages,
including DIL packages, TO-8 and TO-99 cans. The symbol used for the
operational amplifier in circuit diagrams is shown in fig 3.1, along
with the names of the most important connections to the package. The
devices are normally powered by dual supplies of +/- 5 to +/- 15 V,
although some single supply op-amps are also available. Op-amps have
two inputs (an inverting (-) input and a non-inverting (+) input) and
one output, and "operate" on the difference in potential between the
two inputs - producing an output proportional to that potential
difference. An ideal operational amplifier has the following
characteristics:

  a) it produces an output voltage which is directly proportional
     to the voltage difference between the non-inverting and

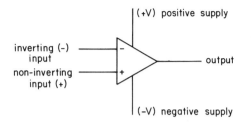

Fig 3.1   The standard circuit symbol for an operational
          amplifier, showing the functions of the essential
          connections. In subsequent figures the power supply
          connections will be omitted.

      inverting inputs.
   b) the constant of proportionality in a), ie. the voltage gain,
      is very large - essentially infinite.
   c) the bandwidth of the amplifier is infinite.
   d) the input impedance at both inputs is infinite.
   e) the output impedance is zero.

      Not surprisingly, available op-amps do not actually achieve these
ideals. Table 3.1 shows the actual properties of some popular op-amps,
and allows us to introduce some of the imperfections of practical
devices which need to be taken into account in the design of
instrumental systems.

      The voltage gain of a practical op-amp is never infinite. Typical
values range from 10^4 to 10^7. These values actually refer to the open-
loop voltage gain, ie the voltage gain of the op-amp when the output
signal is not fed back (or looped) to one of the inputs. In most
applications some part of the output signal will be fed back to the
inverting input, and as we shall see later, this results in the
effective voltage gain of the circuit being quite low (typically 1 -
1000) but stable and controllable. For this reason the fact that the
open loop gain is not as large as infinity does not impose a serious
limitation on the use of op-amps, although the frequency dependence of
the open loop gain is quite a different matter (see below). Of course,
under no circumstances can the value of the output voltage exceed the
supply voltage available to the op-amp; once either the positive or
negative supply voltage is reached the output is said to be saturated.
(Some op-amps cannot provide an output as large as the supply
voltage.)

      The input impedance of practical op-amps is also somewhat lower
than infinity, so that some current is actually drawn from a signal
source connected to an op-amp input. As impedance is a frequency
dependent property, manufacturers normally specify the dc input

Table 3.1   Principal charcteristics of some popular operational amplifiers

Device no.	741	531	725	071	LH0032
Type	- -	bipolar	- -	BIFET	FET
Input resistance	2M	20M	1.5M	10^{12}	10^{12}
Open loop voltage gain / 10^6	0.2	0.06	2.2	0.2	0.003
Input offset voltage /mV	2	2	2	3	2
" temp.co. /microV/°C	5		2	20	25
Input bias current /nA	80	400	80	30pA	5pA
Input offset current /nA	20	50	1.2	3pA	-
" temp.co. /nA/°C	0.5	0.6	0.01	-	
small signal bandwidth /MHz	1	1	0.08	3	70
slew rate /V/microsecond	0.5	35	0.25	13	500
full power bandwidth /kHz	10	500	10	150	7MHz
CMRR /dB	90	100	115	76	60
Maximum output current /mA	20	10	15	10	15

resistance, although for frequencies below 1 MHz this is likely to be adequate for most calculations. For available bipolar op-amps (op-amps constructed using bipolar transistors) the input resistance is typically greater than 1M, although this is not high enough to match the ouput impedance of many important transducers (eg. pH electrodes). Fortunately a type of op-amp which uses field-effect transistors for its input stages (known as a BIFET op-amp) is now readily available, and op-amps in this category can have input resistances of $>10^6$ M.

The ouput impedance of common op-amps is low enough for most purposes, typically being <10 R and sometimes <1 R for dc signals. However, the bandwidth of available operational amplifiers does present something of a problem - not least because it is often difficult to determine precisely over what frequency range an op-amp is likely to function. The relationship between bandwidth and the application of the op-amp compels us to defer discussion until we have examined some op-amp circuits. For the present it will be sufficient to note that while early op-amps were essentially low frequency devices (operating from dc up to about 100 kHz), some of the op-amps

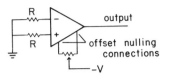

Fig 3.2   An op-amp connected for measurement of input offset current.
          Note that some op-amps require the offset voltage nulling
          potentiometer to be connected to the positive supply rather
          than as shown.

now available will function up to 100 MHz. Sadly those op-amps which
can operate above a few MHz tend to have a number of disadvantages,
such as a high power consumption (causing temperature rises), high
noise levels and a desire to oscillate. Consequently the use of these
devices is desirable only when high frequency operation is essential.
At lower frequencies the more mundane (and much cheaper) op-amps with
modest bandwidths provide far less trouble than their high speed
relations.

     Unfortunately there are a number of other imperfections
associated with practical op-amps, and some of these are sufficiently
important to be considered before we look at the uses of these
devices. Two of the more serious imperfections are called the input
offset voltage and input offset current. These terms refer to
differences between the inverting and non-inverting inputs of the op-
amp. For example, when both input connections are connected together
and grounded, a voltage appears at the op-amp output. The value of
this output voltage divided by the voltage gain of the circuit is
called the input offset voltage. Its value usually sounds quite small
for normal (bipolar) op-amps, typically <2 mV for new designs,
although it can be significantly larger (up to 10 mV) for FET input
op-amps. However, input offset voltage is subject to the same voltage
gain as the difference signal applied between the two inputs, so that
when very small signals are being amplified the input offset voltage
can be an inconvenience. Some operational amplifiers are provided with
two connections which can be used to balance or "offset null" the
input circuits with a single preset potentiometer. An alternative
arrangement which works for any op-amp is considered in section
3.5.3.

     To make matters even worse, the input offset voltage varies with
temperature, the variation being known as the input offset temperature
coefficient or drift, and being specified in mV/C. FET input op-amps
have rather large input offset temperature coefficients, typically
0.02 mV/C. This can be the source of some inconvenience unless some
thought is given to keeping op-amps handling very small signal
voltages away from sources of heat - such as heat sinks of power

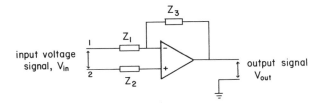

Fig 3.3   Generalised op-amp feedback loop circuit. The op-amp adjusts
          its output in an attempt to make the voltage difference
          between the (+) and (-) inputs zero.

supplies. Furthermore, when an op-amp is driving a low impedance load,
the output current may be large enough to cause heating of the
amplifier package, which in turn can exacerbate drift. To minimise
problems from this source it is preferable to operate high gain
amplifiers with output loads of 10 k or more.

    The second input offset problem is the input offset current. If
the input offset voltage is nulled so that when both inputs are
grounded the output voltage is zero, and the op-amp is then connected
as shown in fig 3.2, it is found that the output voltage is no longer
zero. This arises because the two op-amp inputs draw slightly
different currents, called input bias currents, through the resistors,
R, and so the inputs are at slightly different potentials - thus
giving rise to an output. There is little that can be done to overcome
the effects of the input offset current, beyond selecting an op-amp
which has a small value for this parameter. However, the effect does
remind us that the op-amp inputs do draw bias currents, and that when
these bias currents pass through resistances before reaching the
inputs, the voltage drops across those resistances may contribute to a
voltage difference between the op-amp inputs - thus giving rise to an
output. For this reason it may be important to ensure that the total
resistances of circuits connected to the two inputs of op-amps are
approximately equal.

3.2 Feedback systems

    As indicated above operational amplifiers are not intended to
operate with an actual voltage gain as large as the open loop gain.
Op-amps are provided with such high gains to provide an extremely
versatile building block with which to construct a wide variety of
circuits. In practice operational amplifiers are used in circuits
which feedback some of the op-amp's output to one of its inputs. The
starting point for our brief survey of op-amp applications is the
feedback loop circuit of fig 3.3. In this circuit an input voltage
signal is applied between points 1 and 2. The input bias currents
flowing through the impedances Z_1 and Z_2 to the two inputs of the op-

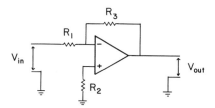

Fig 3.4   The basic inverting amplifier configuration.

amp are usually small, so that the voltage drop across the impedances can be neglected (unless the impedances are very large, >1 M, or the signal very small, <1 mV). Thus the potentials at the inverting and non-inverting inputs may be taken as equal to those at 1 and 2 respectively. The output voltage of the op-amp (with respect to ground), V_{out}, is proportional to $V_{in}=V_2-V_1$, the voltage difference between the inverting and non-inverting inputs. In the absence of the feedback impedance, Z_3, the output voltage would try to reach a value given by the product of the differential input voltage and the open loop voltage gain, G_{ol}, ie.

$$V_{out} = G_{ol}*V_{in}$$

(Of course, it would not succeed if this product exceeded the supply voltage.) However, the presence of Z_3 allows current to flow between the output and the inverting input, and this current opposes that flowing through Z_1 to the inverting input. Because the voltage gain of the op-amp is so large, any finite voltage difference between the inverting and non-inverting inputs would give rise to a saturated output voltage, which in turn would cause current to flow through Z_3 in such a direction as to remove the differential input voltage. The practical result of this is that the output voltage settles to the level required to make the voltage difference between the inputs equal to zero. In the case of fig 3.3 this occurs when

$$V_{out} = V_{in}*(Z_3/Z_1) \qquad\qquad (3.1)$$

While Z_2 does not enter into the equation for the output voltage, we would choose the real part of Z_2 to be equal to the dc resistance seen by the op-amp from the inverting input (ie. the real part of Z_1 in parallel with Z_3) so that the small potential drops associated with the input bias currents were equal at both inputs.

3.3 Basic amplifier configurations

The general principles of the feedback loop considered above are relevant to loops containing any type of feedback impedance. Most straightforward amplification applications involving dc or low

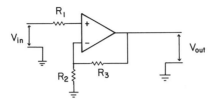

Fig 3.5   The basic non-inverting amplifier configuration.

frequency ac signals require the output signal to be directly
proportional to the applied input signal, and in this case the
feedback elements are purely resistive. There are three basic types of
amplifier circuit which utilise resistive feedback, and these are
illustrated in figs 3.4 - 3.6.

Figure 3.4 shows a schematic diagram for the basic inverting
amplifier configuration. The circuit acts as a voltage amplifier with
a voltage gain of $-R_3/R_1$, (ie. the sign of the output voltage is
different from that of the input signal). R_2 should be chosen so that
its value is approximately $R_1R_3/(R_1+R_3)$ to minimise errors due to
input bias currents. The amplifier is suitable for amplifying signals
ranging from a few mV (positive or negative) to a few volts, subject
to the condition that the amplified output does not exceed the supply
voltage, and is normally used to provide voltage gains of between 1
and 1000, although higher voltage gains can be achieved. While the
resistance values are not critical, it is common practice to use a
value of around 10 k for R_1 with bipolar op-amps, and 100 k - 1 M with
FETs and BIFETs. Higher values can be used, particularly with FET
input op-amps, although there is no escape from the basic limitation
of this configuration - that it has an input impedance characteristic
of the resistance network rather than of the op-amp itself. This
arises because R_2 is grounded on one side, effectively maintaining the
non-inverting input at ground potential, so that the inverting input
potential is always maintained very close to ground (ie. within a few
microvolts) by the action of the feedback loop. In fact the inverting
input of the inverting amplifier configuration is often described as
being a "virtual ground". The applied input signal therefore sees R_1
as connected to ground potential, and the input resistance of the
circuit is R_1. This is generally a much lower value than the
characteristic input impedance of the op-amp, and the inverting
configuration amplifier is therefore inappropriate for the
amplification of signals from high impedance sources.

An alternative configuration which overcomes the problem of a
restricted input impedance is the non-inverting amplifier shown
schematically in fig 3.5. The principal difference between this
configuration and the last is that neither input is now maintained at

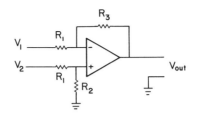

Fig 3.6   The differential amplifier configuration. The differential
          input voltage is (V_2-V_1).

ground potential. The non-inverting input will be at the applied
signal potential, and the inverting input will, by virtue of the
feedback loop, be maintained at virtually the same potential. Thus the
impedance seen by the signal source is R_1 (which is normally
insignificant) in series with the input impedance of the op-amp (which
is typically $>10^7$ R). The signal fed back to the inverting input now
derives from the voltage divider formed by R_2 and R_3, and as this must
equal V_{in}

$$V_{in} = V_{out}*R_2/(R_2+R_3)$$

Rearranging, we find that the ouput voltage is given by

$$V_{out} = V_{in}*(R_2+R_3)/R_2 \qquad (3.2)$$

and has the same sign as the input signal. Again R_1 is chosen to be
approximately $R_2R_3/(R_2+R_3)$ to minimise errors due to bias currents.

While the non-inverting configuration may appear to be an ideal
amplifier circuit, one of the imperfections of practical op-amps now
attempts to spoil an otherwise promising circuit. The same floating of
the inputs which allowed the input impedance to be so high, results in
both inputs essentially following the applied signal. This is fine for
small applied signals of a few mV, but when the applied signal reaches
several volts the output voltage tends to become subject to errors
known as common mode errors, and the gain of the amplifier changes. A
common mode signal is a signal applied to both inverting and non-
inverting inputs (in this case the signal at the inverting input is
derived from the voltage feedback of the feedback loop). For an ideal
op-amp a common mode signal should produce an output of 0 V. The
ability of a practical op-amp to cope with common mode signals is
quantified by the common mode rejection ratio (CMRR), which is the
ratio of the output produced by a normal differential voltage signal
to the output produced by a common mode signal of the same magnitude.
To make matters worse the CMRR is usually quoted in dB and is
typically in the range 60-100, which translates to 10^3-10^5 when

everything is in volts. Thus in an unfavourable case a common mode
signal of 10 V may give rise to the same output voltage as a normal
input difference signal of 10 mV. Possibly not a serious problem for
most dc applications, but a potential source of distortion for ac
signal handling.

Figure 3.6 shows the basic differential amplifier configuration,
an arrangement which is useful because the input signal does not need
to be related to ground. The output voltage in this configuration is
given by

$$V_{out} = (V_2-V_1)*R_3/R_1 \qquad\qquad (3.3)$$

and R_2 is normally chosen to equal R_3 to minimise bias current errors.
This configuration is generally adopted when it is desired to amplify
a small difference signal (V_2-V_1), which may accompany a significant
common mode signal, $(V_1+V_2)/2$, although for most op-amps neither V_1
nor V_2 may exceed the op-amp's supply voltage. The input impedance of
this configuration is high, again because the inputs are not virtual
grounds, but the price paid for this is the risk of common mode
errors. Furthermore, the common mode signal at the op-amp inputs is
maintained by the voltages V_1 and V_2 applied through the two input
resistors R_1, as well as by the contribution of the feedback loop. For
this reason common mode errors are accentuated by errors in the values
of the resistors, and to minimise this source of error closely matched
resistor should be used for the R_1s and for R_2 & R_3. Resistors with
tolerances of better than 1% are normally required, and 0.1% tolerance
would be even better.

3.4 Bandwidth and slew rate

The three basic amplifier configuration of figs 3.4 - 3.6 may be
adopted with most available op-amps and, within the limitations noted,
will perform adequately when amplifiying dc voltage signals. However,
when the input signal possesses a time-dependent component the ability
of the op-amp's output voltage to keep up with changes in the input
signal becomes a major consideration. The relevant details on the
specification of a particular op-amp are contained in two items of
data provided by the manufacturer. The first is a plot of the open
loop gain as a function of signal frequency for small signals. (It is
not always obvious what manufacturers mean by small signals. In
practice it usually means sine waves with amplitudes of a few mV.) A
typical gain vs. frequency plot for a general purpose op-amp is shown
in fig 3.7. Clearly while the open loop gain is large for dc and low
frequency signals, the gain falls dramatically as the signal frequency
rises. The important point is that the closed loop gain of an
amplifier circuit at any frequency cannot exceed the open loop gain of
the op-amp at that frequency. For example, the op-amp specified in fig
3.7 may be used in a circuit operating with a closed loop gain of 100
to amplify small signals with frequencies up to about 10 kHz; but

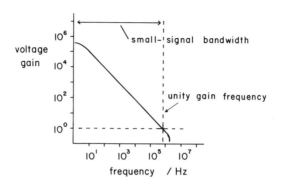

Fig 3.7   Typical example of the variation of the open loop gain of a
           741 op-amp with signal frequency. Such curves may be used in
           designing closed loop systems for handling small signals.
           (Conditions: power supply +/- 15 V, output load 2 k
           resistive, temperature 25 C).

attempts to operate the circuit at higher frequencies will result in a
lower gain than that calculated from the input and feedback
impedances.

      The frequency at which the open loop voltage gain falls to unity
(or 0 dB) is properly referred to as the cut-off frequency, although
it is often more optimistically termed the small signal bandwidth. As
most of the gain vs. frequency plots tend to be straight lines (when
the axes are both logarithmic), the term gain-bandwidth product is
also commonly used to indicate the frequency capabilities of an op-
amp. In our example (fig 3.7) the gain-bandwidth product is
approximately 1 MHz, so that the op-amp may be used at gains of up to
100 for small signals up to 10 kHz, or gains of up to 10 for small
signals up to 100 kHz, or unity gain for small signals up to 1 MHz.

      In practice the gain vs. frequency characteristic for an op-amp
depends on the values of one or two components known as frequency
compensation components. Many op-amps are available in versions which
are "internally compensated", the frequency compensation components
being inside the op-amp package (eg. the 741 is basically a
compensated version of the 748). In these cases the gain vs. frequency
characteristic of the op-amp is fixed. However, uncompensated op-amps
require the user to connect his own frequency compensation components
to pins on the op-amp package, and in this case the gain vs. frequency
characteristic is dependent on the value of these components. The data
sheets for such op-amps usually indicate the manufacturer's
recommendation for compensation components for a variety of
circumstances, so this additional chore is not particulary difficult.
An example of a *10 inverting configuration amplifier circuit complete

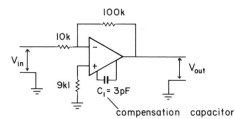

Fig 3.8   Schematic diagram for a *10 inverting amplifier using an
          uncompensated 748 op-amp. The compensation capacitance, C_1,
          is chosen according to the circuit gain and the maximum
          anticipated signal frequency - 1 MHz in this case.

with a single frequency compensation capacitor is given in fig 3.8.

        The second important item of data relevant to the signal
frequency limitations of an op-amp is called the "slew rate". This is
the maximum rate at which the output voltage can change, and is quoted
in volts per microsecond. Values vary widely, as indicated by the
examples given in table 3.1, and may depend on the supply voltage
used. If an op-amp is amplifying a sinusoidal signal which falls
within its small signal bandwidth, and the gain produced by the
feedback loop is increased, then there will come a time at which the
maximum rate of change of the output voltage equals the slew rate for
the device. Beyond this point the amplitude of the output waveform
cannot be increased, either by increasing the gain or by increasing
the input amplitutde. Attempts to produce a slightly larger output may
appear successful when viewed on an oscilloscope, particularly if the
values of the the resistors setting the gain are small (eg. <1 k), but
in reality the output waveform will be distorted and the output
amplitude will not be a linear fuction of the input. The limitation
imposed by the slew rate is expressed in its most severe form by the
full power bandwidth of the device, which is the maximum frequency at
which the output can oscillate with the maximum possible amplitude
(usually close to the supply voltages). The full power bandwidth of an
op-amp may be two orders of magnitude smaller than its small signal
bandwidth, and in general it is only the relatively expensive hybrid
devices which can handle large amplitude signals at frequencies above
a few MHz.

3.5 Practical dc signal circuits

        Now that we have surveyed some of the fundamental characteristics
of operational amplifiers we can turn to a brief examination of some
additional types of op-amp circuits which are useful in the design of
instrumental electronics for the laboratory. The circuits described
below have all been used for specific applications in the author's

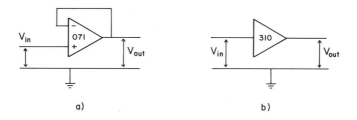

a)   b)

Fig 3.9   a) Schematic diagram for a unity gain voltage follower for
          which $V_{OUT} = V_{IN}$. b) A buffer amplifier with a fixed gain
          (generally unity); such amplifiers do not have inverting
          input connections and so require no external feedback
          components.

laboratory, and so may benefit from some modification before
application in different circumstances. However, it is not the
intention here to provide detailed information on the do-it-yourself
construction of specific circuits, but rather to assist the reader in
understanding the basics of what is involved in various signal
handling aspects of modern laboratory instrumentation.

3.5.1 Unity gain buffer amplifiers

A fairly common requirement in handling signals in a laboratory
is to pass a voltage signal from a high output impedance source (eg. a
transducer) into a circuit of a much lower input impedance (eg. a
signal converter circuit). This may be achieved using a buffer
amplifier which does not change the magnitude of the voltage signal,
but effects only the series impedance presented to a subsequent
circuit (see section 2.5). One of the most useful types of buffer
amplifiers is the voltage follower circuit shown in fig 3.9a. The non-
inverting input accepts the input signal, presenting an input
resistance equal to the characteristic input resistance of the op-amp,
typically $10^6 - 10^{12}$ R. The output connection provides both the output
signal, at the low characteristic output impedance of the op-amp,
typically <10 R, and 100% voltage feedback to the inverting input of
the op-amp. The ouput voltage follows the input voltage, although
offset nulling may be necessary if high precision is required.
Essentially the same effect is produced if resistances are placed in
both the (+) and (−) input lines, and this is often advisable to
protect the inputs. As the circuit is essentially a special case of
the non-inverting amplifier configuration, the voltage follower is
subject to common mode errors. On the other hand the op-amp is
operated at unity gain, so, if only small signals are involved, the
small signal bandwidth of the circuit may be fully utilised.

The use of an internally compensated op-amp allows the voltage

Table 3.2   Typical voltage follower and buffer amplifiers

Device no.	LM302	LM310	LH0033	LH0063
Type	- bipolar -		- - FET - -	
Input resistance	10^{12}	10^{12}	10^{11}	10^{11}
small signal bandwidth /MHz	10	20	100	200
slew rate /V/microsecond	10	30	1500	6000
full power bandwidth /kHz	60	250	-	-
Maximum output current /mA	20	24	100	250

follower circuit to be implemented easily for dc and low frequency signals. However, the design of compensation networks for uncompensated op-amps operating at unity gain and high frequencies is particularly tricky. Fortunately a number of IC manufacturers produce dedicated buffer amplifiers in which the feedback and compensation components are contained in the package, and which are therefore very simple to use. Such buffer amplifiers may have no external inverting input connection, so their circuit symbol appears as a modified op-amp symbol as illustrated in fig 3.9b. Examples of some useful buffer amplifiers for a variety of bandwidths are given in table 3.2.

3.5.2 The intrumentation amplifier

An instrumentation amplifier is essentially a circuit with most of the characteristics of an operational amplifier, but with significantly better specifications in terms of gain, noise, stability, input impedance, CMRR and input offset temperature coefficient. Their principal use is in the amplification of small differential signals coming from high output impedance transducers, particularly in situations where a large common mode signal is present. The basic circuit of an instrumentation amplifier is shown in fig 3.10, where it can be seen that non-inverting amplifier elements are used in each input line to maintain the high input resistance of each input. Offset nulling is provided on one of these stages, as this is all that is necessary to trim the output of the whole circuit.

Several manufacturers produce instrumentation amplifiers in a single package, and for most purposes it is better to buy one of these than to attempt to construct one's own from op-amps. As most of these devices are hybrid circuits they tend to be quite expensive. Nevertheless, if the application really calls for high gain, high precision, high CMRR and low noise, the additional expense can usually be justified. It is perhaps worth pointing out that amplifiers like

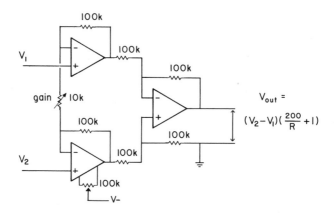

Fig 3.10 Basic circuit of an instrumentation amplifier with overall
voltage gain determined by R, the value set by the 10 k
variable resistor.

the 725, which are called instrumentation operational amplifiers, are
actually good quality op-amps. While they are excellent op-amps for
use in laboratory electronics, they do not have the precision or other
qualities of "true" instrumentation amplifiers - although they don't
cost as much either. Examples of some available instrumentation
amplifiers are given in table 3.3.

Instumentation amplifiers can be used in many of the circuits
presented below, although such extravagance would not normally be
justified. Furthermore instrumentation amplifiers tend to have rather
small bandwidths, being primarily designed for use with dc-producing
transducers such as strain gauges, so their use in ac systems is
somewhat limited. However, they are to be recommended as the
amplifying components of high resolution analog input interfaces as
described in chapter 6.

3.5.3 Summing amplifiers

The existence of a virtual ground at the inverting input of the
inverting amplifier configuration op-amp, allows signals to be
combined in a particulary straightforward manner. Figure 3.11 shows an
inverting configuration amplifier with three independent signal
voltages (V_A, V_B and V_C) applied to the inverting input via the three
input resistors (R_A, R_B and R_C). Bearing in mind that the op-amp in
the inverting configuration will do whatever it can to ensure that its
output applied through the feedback resistor will maintain the
inverting input at a virtual ground, we can see that the current
through the feedback resistor (neglecting the op-amp bias current)
must exactly balance the sum of the three currents flowing through the

Table 3.3   Principal characteristics of some
instrumentation amplifiers

Device no.	LF352	LH0036	LH0038	ICL7605
Type	FET	bipolar	bipolar	FET
Input resistance	$>10^{12}$	300M	5M	–
Gain range V/V	1–1000	1–1000	–	–
Input offset voltage /mV	15	1	0.1	0.005
" temp.coef. /microV/$^{\circ}$C	10	10	0.25	0.1
Input offset current	20pA	10nA	5nA	–
Small signal bandwidth /kHz[a]	7	0.35	1.6	0.01
Slew rate V/microsecond	1	0.3	0.3	0.5
Full power bandwidth /kHz[b]	25	5	–	–
Noise voltage 0.1–10Hz /microV.	1.3	1	0.2	5
Noise current 0.1–10Hz /pA	0.01	–	10	–
CMRR /dB[c]	105	100	114	100

Notes:
a) varies with gain; stated value is frequency at which
nominal gain 1000 drops to 500 (ie. -3dB). b) unity gain
value. c) varies with gain, source impedance and frequency!
quoted value is for gain 1000 and dc.

input resistors, ie.

$$I_{in} = -V_{out}/R_3 = V_A/R_A + V_B/R_B + V_C/R_C$$

Consequently the output voltage is a weighted sum of the three input
voltages, ie.

$$V_{out} = -(V_A R_3/R_A + V_B R_3/R_B + V_C R_3/R_C) \qquad (3.4)$$

This technique is known as "summing" and the inverting input of the
op-amp in fig 3.11 is often referred to as the summing point. If $R_A = R_B = R_C = R_1$, then

$$V_{out} = -(V_A + V_B + V_C)*R_3/R_1 \qquad (3.5)$$

and we have a technique for adding voltage signals together.

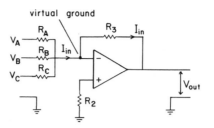

Fig 3.11 A typical summing amplifier. Its operation is made possible
         by the existence of a virtual ground at the op-amp's
         inverting input.

     One useful application of the summing amplifier is in the
provision of an offset or the subtraction of a background level from a
signal, before the balance of the signal is amplified or recorded.
Figure 3.12 shows the circuit of a typical amplifier in which the
input signal may be offset by up to 1 V (positive or negative) before
the signal is amplified by the the *10 amplification stage. Note that
in this example the offset provision is for +/-1 V (before
amplification), and not the +/-10 V applied to the ends of the
potentiometer. This is because the input resistor in the offset input
is 100 k, whereas that in the signal input is 10 k. Thus the input
signal voltage, V_{in}, is effectively multiplied by 10 while the applied
offset voltage is multiplied by 1 (see eqn (3.4)). Lowering the input
resistor in the offset input to 10 k would, of course, allow an offset
of +/- 10 V before amplification, while raising it to 1 M would
provide an effective offset range of +/- 100 mV. If the potentiometer
in fig 3.12 is a multi-turn preset resistor, then the circuit provides
a universal method for offset nulling an inverting amplifier –
although only suitable for the inverting configuration as it relies on
the existence of a virtual ground at the summing point.

3.5.4 The voltage-to-current converter
      (Transconductance amplifier)

     The high input impedance of op-amps makes it convenient for them
to be regarded as voltage amplifiers, and their low output impedance
means that they will maintain a required output voltage over a wide
range of output currents. (Some op-amps are provided with in-built
protection circuits which shut off the output voltage if too much
current is being drawn from the device.) However, there are occasions
when a specific current signal is required under conditions in which
it is not sufficient to rely on the load for that current remaining
constant or at a constant reference potential. An example is provided
by the need to generate a triangular waveform – such as that needed to
modify the source velocity in Mossbauer spectrometry. A
transconductance amplifier allows us to generate a current signal

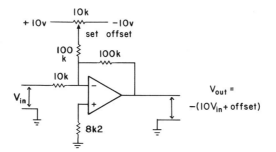

Fig 3.12 Inverting amplifier with a wide range signal offset facility.
The same arrangement may be used to null the op-amp's input
offset voltage.

which can be driven into a varying load or a load operating with a
changing reference voltage.

   The current output of a transconductance amplifier is determined
by a voltage at the input (hence the term voltage to current
converter), and is given by

$$I_{out} = V_{in}g_m$$

where g_m is called the transconductance of the circuit and is in
siemens (ie. reciprocal ohms). The principle of one type of
transconductance amplifier is illustrated in fig 3.13, where it can be
seen that voltage feedback is applied to both inverting and non-
inverting inputs from the ends of the resistor R_m. As usual the op-amp
will adjust its output voltage in whatever way is necessary to ensure
that its two inputs are at the same potential, and if no voltage
signal is being applied (ie. V_{in} = 0), this can only occur when there
is no current flowing through R_m - as only under those conditions are
its two ends at the same potential. When the load (Z_L) is grounded on
one side the zero current condition requires that the op-amp's output
voltage is zero.

   If an input signal, V_{in}, is now applied, the voltage at the
inverting input is different from that at the non-inverting input, so
the op-amp acts to oppose this change by applying the opposite
potential difference across the inputs. This it does by passing a
current, I_{out}, through R_m to produce the required potential
difference, ie.

$$V_{in} = I_{out}R_m$$

So the output current from this configuration (which is independent of
the load Z_L) is

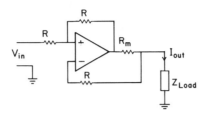

Fig 3.13 One arrangement for the implementation of a transconductance
        amplifier, in which the output current is monitored by the
        voltage drop it produces across R_m.

$$I_{out} = V_{in}/R_m = V_{in}g_m$$

Transconductance amplifiers are available in single packages
(although most of these are based on a different technique from that
described above). An example is provided by the 3080 operational
transconductance amplifier, which allows the transconductance to be
varied by controlling a bias voltage or current at one of the IC's
terminals. The 3080, which has a bandwidth of 2 MHz, may have its
transconductance varied from essentially zero (with a bias current of
zero) up to 9600 microsiemens (with a bias current of about 2 mA.). A
triangular waveform generator based on a 3080 device is illustrated in
fig 3.14, where it may be noted that no external feedback element is
used in this circuit. The waveform is generated by passing a constant
current into a capacitor, the direction of this current alternating
with the sign of a (voltage) square wave applied to the input of the
3080. R_3 is required if the circuit is driving a high impedance load,
as this ensures that the output voltage varies symmetrically about
ground. In the audio frequency range a value of about 100 k should be
satisfactory.

If the load of a transconductance amplifier is a resistance, R_L,
then the output voltage (ie. across the load) is

$$V_{out} = R_L I_{out} = V_{in}g_m R_L$$

so that the circuit behaves essentially as a voltage amplifier (in
fact the 3080 offers quite a useful bandwidth considering the low cost
of the device), but this voltage amplifier offers a gain which may be
varied simply be varying the bias supply. Furthermore the gain can be
changed from any particular value to zero and back - again by
switching the bias supply - so that the signal being amplified may be
gated, ie. passed according to the level of the (gating) signal on the
amplifier's bias pin. Until recently these were valuable techniques to
apply in the design of interfaces for microcomputers, although the

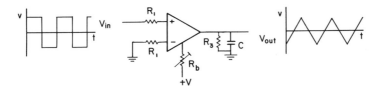

Fig 3.14 A square-to-triangular waveform converter based on an
         operational transconductance amplifier, which produces an
         output current used to charge the capacitor C.

improvements in the quality and speed of analog switches and
programmable op-amps in the last few years have partly eclipsed the
value of the transconductance amplifier in this role.

    Operational transconductance amplifiers are relatively low-power
devices capable of rather small output currents - typically a few
microamperes - when compared with normal op-amp outputs, and require
the addition of current amplifiers (eg. bipolar transistors) to
generate higher currents. However, transconductance circuits based on
op-amps and typified by fig 3.13 are also useful when substantial
current signals are required or where higher voltage ranges than the
normal +/- 15 V need to be covered, as high output current or high
voltage op-amp types may be used.

3.5.5 The current-to-voltage converter
      (Transresistance amplifier)

    A transresistance amplifier produces a voltage output
proportional to a current input. The applications of transresistance
amplifiers may be more immediately apparent than those of
transconductance circuits, because a number of important transducer
devices (such as the photomultiplier tube) produce their output in the
form of a current signal which needs to be converted into a voltage
signal before its level can be shifted, amplified or recorded. A
typical op-amp based transresistance amplifier is shown in fig 3.15.
In this example the input is the current from a photomultiplier's
anode, and it is worth noting that there is no other connection to the
tube's anode. The tube is, of course, operating with a negative
photocathode; this avoids the use of high voltage blocking capacitors
in the signal lead and minimises interference from the high voltage
supply and cabling. The anode is maintained at the required ground
potential by the virtual ground at the inverting input of the op-amp,
which in turn is present because the non-inverting input is grounded
through R_2. To maintain the virtual ground the op-amp output voltage
takes the value which causes the current in the feedback resistance,
R_3, to be equal but opposite to that from the anode. Thus

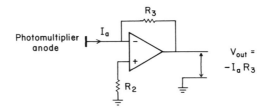

Fig 3.15 A transresistance amplifier used to generate a voltage signa
        from the anode current of a photomultiplier tube.

$$V_{out} = -I_a R_3$$

The value of R_3 is chosen to produce a reasonable voltage output for
the expected anode current; in our case we selected 10 V output for
the maximum anticipated anode current of 0.1mA, which required a
feedback resistance of 100 k and allowed our normal signal levels to
be around 1 V. R_2 was chosen to equal R_3 as the latter is the only
hardwired pathway for the input bias current to the inverting input.

3.6 Ac signal circuits

The circuits discussed above will handle ac voltage and current
signals in addition to constant level signals, and often there is no
need to worry about changing an amplifier's circuit when moving from
dc to an ac signal. However, there may be some advantages to handling
an ac signal in a different way, and sometimes there is a need to
handle ac signals unhindered by any superimposed dc level.

3.6.1 Ac amplifiers

Voltage and current amplifiers for ac-only signals can be
assembled simply by including a capacitor in the input signal pathway
of a dc amplifier. Such capacitors may be called dc-blocking
capacitors or ac-coupling capacitors – depending on your point of
view. Of course it is important to choose the value of the capacitanc
so that it provides a low impedance at the signal frequencies of
interest (relative to the impedance of any gain setting resistors).
Figure 3.16 illustrates two points of interest in adapting
straightforward amplifier circuits to handle ac-only signals. The
particular example is for an audio frequency range, non-inverting
configuration ac voltage amplifier, but the principles apply to all
ac-only systems (of course, the choice of inverting or non-inverting
in the ac-only case depends on the desired circuit characteristics
such as stability and input impedance, as the inversion of an ac
signal effects only its phase).

The first point is that if the dc pathway to either op-amp input

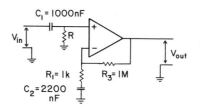

Fig 3.16 A *1000 ac voltage amplifier circuit which provides for a
         gain roll-off for frequencies below 10 Hz to minimise dc
         errors and drifts.

is blocked by a capacitor then some alternative route for the input
bias current must be provided. In our example this route is provided
by the addition of the resistance R between the non-inverting input
and ground. This resistor does load the input signal and, by strapping
the non-inverting input to ground, ensures that a (time averaged)
virtual ground exists at the inverting input. (this is useful if
shifting of the mean signal level is required as an offset may be
applied at the non-inverting input.) The value of R may be selected in
the normal way (ie. to be equal to the effective bias current pathway
resistance to the inverting input), but note that R_1 does not provide
a dc pathway to ground because of the blocking action of C_2. However,
this consideration is not important if this circuit's output is ac
coupled, because any dc offset errors will be blocked by the coupling
capacitor. The input circuit made up of C_1 and R also forms a high-
pass filter, with a corner frequency equal to $C_1 R$, about 10 Hz with
values of R = 100 k and C_1 = 1000 nF. Thus the input signal is
attenuated at lower frequencies at the rate of 6 dB per octave.

   The second point concerns the inclusion of the capacitor C_2. One
of the most frustrating forms of nuisance on a signal being recorded
on, say, a chart recorder is the slow drift which results from an
apparatus warming up over a period of several hours. This phenomenon
can arise from temperature changes on the circuit board altering the
input offset voltages of op-amps (see the input offset temperature
coefficient, section 3.1), and is hard to control where the op-amps
must amplify dc levels with a high gain. However, when ac-only signals
are involved it is possible to roll-off the gain of an amplifier using
the trick shown in fig 3.16. At zero frequency C_2 has an impedance of
infinity, so 100% of the op-amps output signal level is fed back to
the inverting input and the amplifier has unity gain. At high
frequencies the impedance of C_2 is much smaller than R_1, so the amount
of feedback falls and the gain rises to $(R_1+R_3)/R_1$. With the values
given in fig 3.16 the high frequency gain is 1000. The corner
frequency is that at which the impedance of C_2 = 1 k, making the gain
half of its higher frequency value, ie. at approximately 8 Hz in our

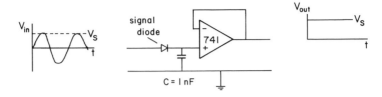

Fig 3.17 The simplest technique for ac-dc conversion. It works over a
wide range of frequencies but results in errors because of
the diode drop.

example. Thus signals in the frequency range of interest (100 Hz - 10
kHz) are amplified 1000-fold, while low frequency drifts and op-amp
offsets are amplified by 1, or less for very low frequency noise on
the input signal which has already been attentuated by C_1R.

3.6.2 Ac-to-dc conversion

While it is convenient to handle ac signals for purposes of
amplification, recording the value of such signals by chart recorder
or computer requires that an ac - dc conversion takes place. In some
cases it is adequate to rectify an ac signal by traditional methods -
such as passing the signal through a diode followed by low-pass
filtering to remove the residual ripple, as illustrated in fig 3.17.
In this example the circuit's input signal should derive from a fairl
low impedance source to avoid excessive loading during the positive
half cycles resulting in errors. The output is buffered using a
voltage follower to ensure that the effective loading of C is small,
so that the dc output records the peak ac input voltage less a "diode
drop" (typically 0.6 V for silicon diodes and 0.2 V for germanium
ones). This circuit will function up to high frequencies (eg. 50 MHz,
if you take care to avoid stray capacitance around the input) and its
response time (ie. how rapidly it can follow changes in ac amplitude)
is give by CR^{-1} s, where R is the input impedance of the op-amp plus
the backwards resistance of the diode (ie. several M). With the value
given the response time is around 10 ms. If this is too slow for a
particular application then a load resistor can be added in parallel
with C.

The main problem with this kind of circuit is that the diode
characteristic is distinctly non-linear, conduction through the diode
producing a voltage drop across it. Although there are special types
of diodes which can help in this situation, an alternative technique
is to overcome the diode drop using an op-amp.

Figure 3.18 shows a typical "precision rectifier" circuit,
capable of providing a dc output voltage equal to the amplitude of th
ac input voltage over the range of 1 mV - 5 V and 1 kHz - 1 MHz (in

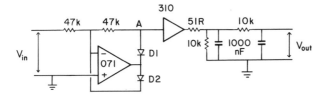

Fig 3.18 A precision rectifier for ac-dc conversion of small signals
         without diode drop errors. Works for signal frequencies up to
         about 1 MHz with the op-amp shown.

this case the signal does not have to be an ac-only signal). For
positive input half-cycles D1 conducts and D2 is "off", so the circuit
behaves as an inverting amplifier with unity gain as far as the signal
at the first stage output (point A) is concerned - although at the op-
amp output the voltage stays a diode drop below the input level. For
negative input half-cycles D1 is off while D2 conducts, forming an
inverter through D2 and allowing the op-amp's output to stay a diode
drop above ground (remember the inverting input is a virtual ground
when the non-inverting input is grounded). With the op-amp's output
positive there is no signal passage through D1 to the buffer. The
buffer is required in this type of circuit because the output
impedance of the first stage (ie. at point A) alternates between a low
value (when D1 is conducting) and 47 k (when D1 is off and the virtual
ground level (the inverting input) is seen through the 47 k resistor).
The buffer presents a high input impedance to the half-wave signal, so
it doesn't load the signal at either impedance level, and provides a
low impedance output to drive the low-pass filter.

    The circuit as shown in fig 3.18 will handle small signals at
frequencies up to about 1 MHz when germaniun diodes are used. For
silicon diodes the greater diode drop voltage changes required as the
signal passes through zero tends to restrict the application to
frequencies below a few hundred kHz, because of slew rate limitations.
For larger signal amplitudes slew rate limiting restricts the output
amplitude in either case to a few tens of kHz.

3.7 Integrators

    Important classes of circuits for instrumentation are those which
enable a signal to be either differentiated or integrated. The
differences between such circuits and those we have met already are
actually quite small, although the effect produced on a signal may be
large. An integrator is a circuit which produces an output voltage
given by

$$V_{out} = \int V_{in} \cdot dt$$

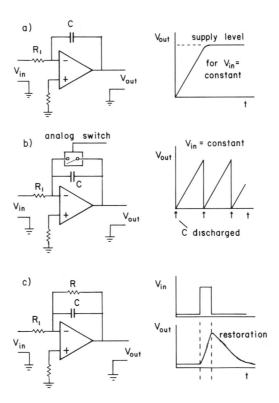

Fig 3.19 The basic integrator circuit (a), which may be modified (b)
to provide for a reset of the output to zero, or (c) to allo
the output to be restored to zero with a fixed time constant
given by CR.

This function is achieved in principle by a circuit such as that show
in fig 3.19a. Here the op-amp's feedback element is a capacitor which
charges up as the output voltage changes to pump a current equal to
V_{in}/R_1 into the capacitor. This results in an output voltage given by

$$V_{out} = (CR_1)^{-1} \int V_{in}.dt$$

and CR_1 is termed the integration time constant. An example of the
integrated output resulting from a constant V_{in} is also shown. Note
that the slope of the output voltage is proportional to $(CR_1)^{-1}$, so
that large integration times give a slower response.

An immediate problem arises with this circuit because its
integrating action cannot be turned off without turning the power off
Two alternative circuits are useful in providing a practical
integrator, the choice between them depending largely on the nature o

the application and the timescale required for the integration. In the
first circuit, fig 3.19b, a switch is provided to short out the
capacitor and reset the integrator output to zero. This switch could
be a manually operated switch, although this could result in the
collection of interference signals which are likely to be picked up on
wires between components and case-mounted switches. A better choice
would be an FET or analog switch in which the signal remains close to
the circuit board and the switch is operated by a simple control
voltage applied to the FET gate or the IC's open/close pin. (Analog
switches are IC electronic switches which have essentially infinite
resistance $(>10^{12}$ R) when the open/close pin is at ground potential
and a very low resistance (10 - 100 R, depending on the switch type)
when the open/close pin is at the positive supply voltage.)

This type of circuit is useful when integration is to be
performed over periods of minutes, and can also be used for periods of
seconds if the switch is operated automatically by the application of
timed open/close signals. Typical applications are in the measurement
of heat output (eg. calorimetry) or light output (eg. thermo-
luminescence dosimetry) where the input voltage signal is provided by
a suitable transducer. Incidentally the circuit can be used for
current integration by omitting the input resistor. In any event the
most likely difficulty with this kind of circuit arises from the
inclusion of offset and bias currents in the integrated output,
although the choice of an FET input op-amp helps to minimise this
problem. For long-time integrations the use of a microcomputer to
numerically integrate a recorded signal voltage is probably a more
versatile and reliable alternative.

The second practical integration circuit, shown in fig 3.19c,
avoids the output drift of the previous circuit by the addition of a
restoring resistor, R, across the feeback capacitor. This allows the
voltage across the capacitor to be restored slowly to zero, with a
time constant given by CR and termed the restoration time constant.
Although there is no reason why large value resistors (or other tricks
to simulate them) should not be used, most applications of this
technique are associated with relatively short time scale
integrations, such as the integration of a current pulse from a
nuclear particle detector or photomultiplier tube.

Without R_1 in the input line the circuit of fig 3.19c becomes a
current integrator. As the integral of current is charge, this type of
circuit is also called a charge sensitive amplifier, giving a (peak)
output voltage proportional to the charge collected at the input
during the integration time, ie.

$$V_{out} = C^{-1} \int I_{in}.dt = Q_{in}/C$$

Note the difference between this approach and that based on the use of
a voltage amplifier operating on the voltage signal produced when the

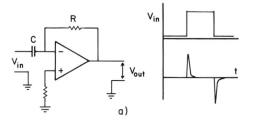

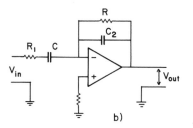

Fig 3.20 (a)The basic differentiator circuit. (b) A differentiator
        with a high frequency roll-off to limit high frequency noise
        and spurious oscillation. This circuit forms the basis of a
        pulse shaping amplifier.

charge collected by a charge gathering transducer effects the
electrode's potential through $V = Q/C_T$ (section 2.1). The charge
sensitive amplifier produces an output which is largely independent of
the transducer's capactiance C_T. In this case it is the maximum value
of the integrated output signal which provides the required
information, and this is proportional to C^{-1}. The restoration time
constant may be chosen to be as large as possible while avoiding the
overlap of adjacent signals.

3.8 Differentiators

    The basic differentiator circuit is given in fig 3.20a and
produces an output voltage given by

$$V_{out} = -RC(dV_{in}/dt)$$

In this case the feedback element is a resistance and the input
voltage signal is coupled to the op-amp's input by a capacitor. An
example of a differentiated output produced by a pulse input is shown
in the figure, although the form of the output would depend on the
pulse duration and the component values used in the circuit. The time
constant given by the product RC is called the differentiation time
constant, and determines the magnitude of the output voltage for a

given rate of change of input signal. As the feedback element in the
differentiator provides a dc pathway for the op-amp's bias current,
differentiators do not suffer from the drift problems of simple
integrators, although the basic circuit shown does produce an output
which can be distorted by contributions from high frequency noise and
interference. Furthermore differentiators which are not frequency
limited are prone to high frequency oscillation. The addition of
frequency limiting components, which cause the gain to roll off at
high frequencies, makes the circuit, shown in fig 3.20b, look similar
to that of the integrator in fig 3.19c. The resistor R_1 in the input
line limits the gain of the circuit by providing a upper limit of
R/R_1 at high frequencies (where the impedance of C has become small),
while the capacitor C_2 ensures that at high frequencies the circuit
behaves as an integrator with an integration time given by R_1C_2 s.

Finally it is worth bearing in mind that pulses form a class of
signals which inconveniently fall between the realms of dc and ac.
They often require handling circuits which have high bandwidths, so
they sound like good candidates for amplification by ac circuits with
capacitors to pass the signal from one stage to the next. However,
passing a unipolar pulse through a capacitor to a load resistor
results in at least partial differentiation of the signal. For this
reason it may be better to differentiate the input pulse in a precise
manner and produce a bipolar signal pulse in the first place, as these
pseudo ac signals are less susceptible to distortion through
interactions with stray circuit impedances than are unipolar signals.
Also bipolar signals could then be handled by ac-coupled circuits,
which tend to have fewer stability and offset problems than their dc-
coupled counterparts. On the other hand ac-coupled circuits do have
more complex frequency dependences of their transfer functions, so
that if a wide range of pulse shapes needs to be accomodated then the
choice is more likely to fall to the dc-coupled system.

3.9 Pulse amplifiers

Pulse amplifiers are essentially refined versions of the type of
circuit shown in fig 3.20b. Many commercial pulse amplifiers,
particularly those designed for nuclear instrumentation, are provided
with switch selectable time constants for both differentiation and
integration. The choice of these time constants depends on the
information desired from the pulses, and the effects of various
combinations of time constants on the output signals generated from a
standard input "step" (an abrupt change in input voltage usually
adopted as an extreme model of the long tail pulse of fig 2.3d) are
illustrated in fig 3.21. If the magnitude of the pulse height is
desired (eg. for pulse height analysis) then this may best be measured
using a short integration time constant (comparable to the rise time
of the incoming signal), as this allows the most accurate pulse height
conversion. Similarly if a long differentiation time constant is
chosen then the peak of the output signal decays more slowly and is

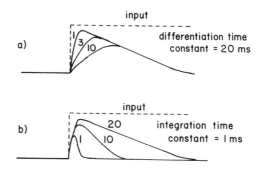

Fig 3.21 Outputs of a pulse amplifier for an input voltage step,
         showing the effects of varying (a) the integration time
         constant while the differentiation time is fixed, and (b)
         vice versa. The value of the varying time constant is shown
         beside each trace.

therefore easier to measure accurately. However, against this one must
note that with a long differentiation time constant the output pulses
may overlap one another, making the accurate measurement of signal
pulse heights less reliable because of pulse "pile up".

    On the other hand, if fast pulse counting is the objective then
preserving the pulse height is not as important as ensuring that the
pulses are not missed by accidental overlapping, so short
differentiation time constants are attractive, with the integration
time constant chosen to maximise the pulse height. Unfortunately short
integration time constants will allow fast noise pulses to be
recorded. Sadly there is no infallible guide to how one should select
these time constants, and the choice will depend on the application
and the shapes of signal and noise pulses. A useful starting point is
usually to match the time constants to the rise and fall times of the
input signals. However, it is important to realise that the selection
of different time constants does have a profound effect on the shape
of the output signal. In fact, by carefully selecting the time
constants it is possible to discriminate against some shapes of input
pulses while amplifying others, and this forms the basis of one
technique of distinguishing between different types of ionising
radiation detected by nuclear radiation detectors.

3.10 Filters

    Filters are circuits which have transfer functions which allow
the passage of signals of one frequency, or range of frequencies,
while blocking those of other frequencies. Like most of the topics
covered in this text, the subject of filter design is very large and
we can do no more that examine briefly one or two examples of

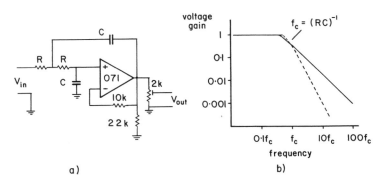

a)   b)

Fig 3.22 A two pole low pass filter (a) giving a frequency response
(b) with a corner frequency, f_c, determined by the two RC
networks. The broken line shows the response obtained by
connecting two filters in series.

important types of filters. We will consider examples of active filter
circuits (ie. those containing at least one powered device capable of
amplification), although filter circuits can be constructed using only
passive (unpowered) components if a loss in signal amplitude is
acceptable.

The first example is a low pass filter - a circuit which passes
signals of frequencies lower than the circuit's corner frequency, f_c,
while attenuating signals of higher frequencies. The circuit is shown
in fig 3.22a and its response to ac signals as a function of frequency
is in fig 3.22b. The filtering action is brought about by the RC
networks in the op-amp's non-inverting input line and feedback loop -
two RC units, and its called a two pole filter. Because the two RC
networks have the same component values the response of this filter is
known as a Butterworth characteristic, and above the corner frequency
of RC^{-1} the gain declines at 12dB per octave (in terms of voltage gain
that's a factor of about 4 for each doubling of frequency). By
selecting the two RC sections to have different component values,
other rates of gain decline may be chosen, Chebyshev (faster decline)
and Bessel (slower decline) being commonly used filter characteristics
although having a more variable low frequency gain than the
Butterworth type.

The preset variable resistor providing the output from the filter
circuit of fig 3.22a may be adjusted to produce an overall gain of
unity for low frequency signals (without that facility the voltage
gain is actually about 1.6 when the Rs are 10 k) and these circuits
may be cascaded to produce a sharper cut-off above the corner
frequency. Low pass filters are particulary useful for the attenuation
of high frequency noise accompanying dc or low frequency signals
produced by laboratory transducers. A corner frequency of less than a

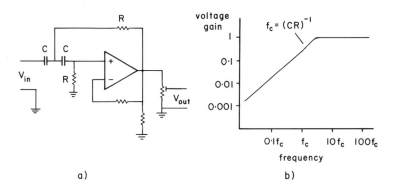

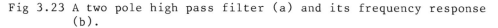

a)   b)

Fig 3.23 A two pole high pass filter (a) and its frequency response
        (b).

few hertz enables dc signals to be cleaned-up by the removal of
interference, such as that derived from the 50 or 60 Hz line
frequency.

    Interchanging the Rs and Cs of fig 3.22a produces a high-pass
filter of the type shown in fig 3.23a and this has a response of the
form shown in fig 3.23b, allowing the passage of signals with a
frequency above the corner frequency, while attenuating dc and low
frequency signals. Again different filter characteristics can be
obtained by selection of the component values, although of course in
this case an upper frequency limit to the transmitted signal is
imposed by the bandwidth and slew-rate of the op-amp used. Combining
low and high pass filters with overlapping pass frequency ranges gives
the band-pass filter, which passes a signal within a given frequency
range (the pass band) while attenuating signals outside that range
(both lower and higher frequencies). In much the same way summing the
outputs from non-overlapping low and high pass filters can be used to
generate band-stop filters. However, it is not so easy to cascade
band-pass and band-stop filters because of the sensitivity of the band
frequencies to the precise values of the components in the RC
networks.

    While the filter circuits we have considered have been based on
conventional op-amps, each type of filter characteristic may be
obtained from purpose designed ICs known as state-variable filters.
These enable high quality filters of each type to be constructed with
a minimum of difficulty and only a small number of additional
components. Furthermore, the manufacturers usually supply detailed
tables of component values required for the different filter
characteristics and for a wide range of frequencies extending from Hz
to MHz. A typical low cost and versatile device is the MF10C, which
can be used with signal frequencies up to about 200kHz, although with

corner frequencies only as high as 20kHz. This device may be
configured for high, low or band pass, and for Butterworth, Bessel,
Cauer and Chebychev characteristics.

CHAPTER 4

THE ELEMENTS OF DIGITAL SIGNAL HANDLING

By comparison with analog signals we should find digital signals quite straightforward, because they lack the variety of forms which plague the analog world. A digital signal can have only one of two possible values on any signal carrying conductor at any given time. The two possible values are known as logic levels, and for the purposes of generalising are referred to as "high" and "low" levels (more commonly abbreviated to hi and lo). The simplest electronic components designed specifically for handling these signals are called logic gates, and there are hundreds of gates on the market – each designed to produce one or more outputs whose level depends on the level applied to one or more inputs. To utilise these circuits it is necessary to define electrical signal values to represent the levels of hi and lo. One of the most widely used conventions requires the level hi to be represented by a voltage of +5 V, and the level lo by 0 V, both being relative to ground. Later in this chapter we shall examine this convention a little more closely and consider one or two alternatives.

4.1 Logic gates

Figure 4.1 illustrates the schematic representations (the symbols) used for two types of logic gate. In fig 4.1a the two input AND gate is shown. The output level from this gate is hi when the levels applied to input a and input b are both hi. If either or both of the inputs are lo, then the output is lo. The two input OR gate is illustrated in fig 4.1b. In this case the output of the gate is hi if either input a or input b is hi. If both inputs are lo then the output is low. If both inputs are hi then the situation may be regarded as meeting the requirements for a hi output, and this gate is more properly termed an inclusive-OR gate (ie. its operation is inclusive of the two inputs hi condition for hi output). An alternative gate, called an exclusive-OR (EOR) gate, treats the two input hi condition as not meeting the requirements for output hi, and gives a lo output in that situation.

Describing the action of gates is greatly simplified by the use

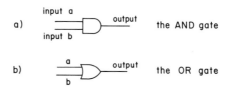

Fig 4.1   Symbols used in circuit diagrams to represent AND and OR
          logic gates. The devices also require a power supply and a
          ground connection, but these are not usually shown.

of a logic symbolism, and this has the additional advantage of
introducing a digital significance of the signal levels. The symbolism
allows us to use 1 and 0 or "true" and "false" (after Boole) for the
signal levels hi and lo. By far the most widely used logic system is
that in which a 1 represents the hi level and a 0 represents the lo
level, and this is the logic system used by manufacturers in naming
the functions of their gates. Naturally the alternative is also used,
and in the context of this book this alternative appears in a number
of important applications, (for example, the GPIB in chapter 7). The
system with hi = 1 is called positive logic, or sometimes hi=1 or
high-true logic. We shall use this system wherever possible. The
alternative system which uses hi = 0 is termed the negative logic
system. The difference between the two systems is of major importance,
as may be seen by writing out the symbolic representation of the
operation of the AND gate using both positive and negative logic as
shown in table 4.1. This type of representation is known as a truth-
table and is the principal description of the operation of a gate in
manufacturers' data sheets. The negative logic output for this gate

Table 4.1   Operation table for a two-input AND gate and
            associated truth tables

electrical operation			positive logic			negative logic		
a	b	output	a	b	output	a	b	output
lo	lo	lo	0	0	0	1	1	1
lo	hi	lo	0	1	0	1	0	1
hi	lo	lo	1	0	0	0	1	1
hi	hi	hi	1	1	1	0	0	0

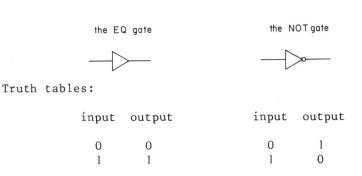

Truth tables:

	the EQ gate			the NOT gate	
input	output			input	output
0	0			0	1
1	1			1	0

Fig 4.2   Symbols and truth tables for the gates providing the EQ
(buffer) and NOT (inverter) functions.

symbolises a different function from the positive logic output, as in
one case different inputs give the same output as two 0s, and in the
other case different inputs give the same output as two 1s. In fact
the function achieved in the negative logic section of table 4.1 is
not the AND function (output 1 only when both inputs 1) but the
inclusive-OR function (ouput 1 when either input 1). (Of course the
gate doesn't know the difference, it always produces a hi output only
when both inputs are hi.)

Two other important gate functions are the EQ and the NOT
functions, although the gates which implement these functions alone
are commonly called buffer and inverter gates respectively. The gate
symbols and truth-tables for these functions are given in fig 4.2.
Note that the difference between the two symbols is the presence of
the small circle on the output of the NOT function gate. The NOT
function is most commonly encountered in combination with some other
function, such as AND or OR, and gates which implement the combined
functions are probably the most widely used of all logic gates. The
symbols and truth table for the NOT AND combination, abbreviated to
NAND, and the NOT OR combination, abbreviated NOR, are shown in fig
4.3. Note again the presence of the small circle on the output lines,
indicating the addition of a NOT function to the function symbolised
for the gate.

The real power of circuits made from logic gates lies in the ways
in which the individual gates may be combined to produce ouputs whose
levels depend on the levels of several different inputs. We can
demonstrate this in a small way by examining the circuit shown in fig
4.4. This is designed to react in a precisely defined way to the
action of three switches: one used for sensing whether a door is open
or closed; another operated by a relay connected to the power-on
cicruit of an X-ray generator; and a third used as a reset switch. In
operation the circuit could be used to activate an alarm (which could

the NAND gate   the NOR gate

=⊐D○—   =⊐D○—

Truth tables:

a	b	output		a	b	output
0	0	1		0	0	1
0	1	1		0	1	0
1	0	1		1	0	0
1	1	0		1	1	0

Fig 4.3 Symbols and truth tables (positive logic) for the gates
         providing the two input NAND and NOR functions.

be a bell, or a light or a relay to turn the power off) if the door of
a safety enclosure of an X-ray diffractometer is opened while the
power is turned on to the X-ray generator. The alarm does not turn off
again even if the door is closed or the X-ray power turned off, unless
a reset switch is operated. Furthermore the alarm operates in the
event of a breakage in the wires transmitting the signal from either
the door swich or the relay switch. (Note: this circuit is provided as
an example of the use of logic gates and not as an example of safety
systems. The author does not claim to be qualified in the design of
safety systems.)

     For simplicity we shall assume that the door switch provides a 0
signal (ie. 0 V) when the door is closed, and that the relay provides
a 0 signal when the generator power is off. If either switch opens, or
the relevant wire is broken, then the input to the gates, G1 or G2, is
taken to a 1 (eg. +5 V) by the 2 k "pull up" resistors. Gates G1 and
G2 are inverters and output a value which is the complement of the
input signal. G3 is a NOR gate and provides an output of 1 only when
both inputs are 0s, ie. when either the power to the generator is on
and the door is open, or when one of those conditions has been met and
there is a fault in the other input channel resulting in a 0 level at
G3's input. Thus the output of G3 is a 0 under normal conditions but
becomes a 1 when the alarm is required to operate.

     If the alarm was connected at the output of G3 then it would
operate under the required condition (door open and power on), but it
would stop operating if the door was closed again. We choose to
complicate matters a little further (and to illustrate a useful
technique) by insisting that the alarm, once activated, continues to
operate until a reset switch is toggled. Gate G4 is a NOR gate

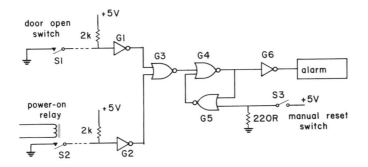

Fig 4.4   Example of a simple logic system constructed from gates. S1
           is closed when the door is closed, and S2 is closed when the
           generator power is off. The reset switch (S3) would be a
           momentary action, normally-open switch.

accepting one input from the sensing circuit (ie. G3) and another from
the output of a second NOR gate, G5, which has one of its inputs
connected to G4's output and the other connected to a normally-
grounded line from the reset switch. Consider first the situation in
which G3's output is a 0 (ie. no alarm) and G5's output is a 0. Both
inputs to G4 are 0s, so G4's output is a 1. Thus one input to G5 is a
1 and the other is a 0 (being held at ground by the 220 R resistor
until the reset switch is operated). Under these conditions the output
of G5 is a 0, so providing G4 with its second 0 input.

On the other hand when an alarm condition is sensed the output of
G3 becomes a 1, so the output of G4 becomes a 0, the output of G5
becomes a 1, and this ensures that the output of G4 remains a 0 even
when the output of G3 reverts to a 0 (ie. the alarm condition is
removed). Only when the normally grounded input to G5 is switched to a
1 (the 5 V signal connected to the other pole of the switch) does the
output of G5 become a 0, allowing the two 0 inputs to G4 to produce a
1 from G4's output again. Finally the output of G4 is connected to an
inverter, G6, which produces the 5V output to operate the alarm only
when its input is a 0 (at this stage we assume that the alarm operates
on receipt of a 5 V signal level, and we refrain from considering the
amount of current it may need).

It is worth spending some time ensuring that the operation of the
circuit in fig 4.4 is understood, as it illustrates not only the
connection of logic gates to produce a circuit which behaves in a more
complex, but still logical, manner, but also the important technique
of the "hold until reset" circuit. As an exercise you may like to try
to improve the circuit, by reducing the number of gates used.

4.2 TTL families

Before we move on to consider more sophisticated logic functions we will examine briefly some of the commercial implementations of the simple gates we have met so far. All commercial gates are available as integrated circuits - at this level they are called small scale integration, SSI, circuits - packaged in DIL packages having 14 connections. As this is a larger number of connections than required for the one and two input gates we have considered above, it is not surprising that most of the 14 pin DIL packages actucally contain several gates, typically four two-input NAND gates etc. Thus in fig 4.4 the gates G3 - G5 could all be in the same package, as could G1, G2 and G6. If you had replaced G1-G3 with a single AND gate the circuit would actually have required more packages than it does as shown. (What could you do to lower the package count?)

There are several different "families" of gates available, although we shall consider only two classes of these. The first class is that based on circuitry known as transistor-transistor logic (TTL) of which the oldest family is that known as the 74 family. Some examples of 74 family packages are listed in table 4.2, showing the basic numbering system which has been adopted by all major manufacturers and inherited by the newer families we shall discuss below. All the TTL families operate with supply voltages of +5 V and ground, but there are other characteristics which are different for the different families. There are four major factors which determine the suitability of a family for any particular application:

1. The signal levels used to represent the hi and lo levels.
2. The speed with which gates can change their signal levels.
3. The number of inputs which may be directly connected to a single output (known as the fan-out).
4. The power required to operate each gate.

For the 74 family (and in fact for all the TTL families) the signal levels are voltage ranges, with fairly stringent current requirements which cannot be neglected. The hi level for the 74 family is the a positive voltage in the range 2.4-5 V, although 3.5 V is a typical value found for the hi ouput of a gate. This output is not intended to be a source of current for driving low impedance devices such as indicator lamps, and in fact the current available from hi outputs is quite small. The lo level is a voltage between 0 and 0.8 V, although gates providing a lo output produce a voltage which is within a few tenths of a volt of ground. One important characteristic of the lo state is that as an output level it is capable of sinking up to about 16 mA of current (ie it will absorb up to 16 mA from a positive potential source) in an attempt to achieve the required output voltage. (So in fig 4.4 it would have been better to operate the alarm with a 0 output rather than a 1.) To produce a lo level at the input

Table 4.2   Examples of 74 series packages of gates

code number	no. of devices per package	type of device
7400	4	2-input NAND
7401	4	2-input NAND oc
7404	6	inverters
7406	6	inverters oc
7409	4	2-input AND oc
7410	3	3-input NAND
7420	2	4-input NAND
7430	1	8-input NAND
7402	4	2-input NOR
7408	4	2-input AND
7432	4	2-input OR
7486	4	2-input EOR

note: oc = open collector outputs

of a 74 family device we need to be able to sink about 1.6 mA to ground (whether this is achieved by the output of a previous gate or by some other means). As a result of the requirements of the 74 family lo level, the fan out of the 74 family is 10, ie. we may connect up to 10 gate inputs to a single gate output. Of course two outputs should never be directly connected together, unless they are special gates known as "open collector output" gates. These require a resistor to be placed between the output and a hi point. When a lo level is to be produced the open collector output drops to ground, sinking current through the resistor, but otherwise an effective hi output level is maintained by the hi point and the gate output draws no current. Typically a resistor of 1-2 k is used for "pulling up" an open collector output, so that it can provide current for other TTL inputs, although if the output is not driving TTL a higher value resistor may be appropriate.

One important point associated with the outputs of the 74 family devices is that they generate a large current spike on the power supply line whenever a gate's output changes level. Thus a good quality bypass capacitor is required as close as possible to the power supply connection to every 74 family package on a circuit board. Omission of this small detail can give rise to the most chaotic circuit behaviour, particularly when other types of IC are present on the same circuit board.

Fig 4.5 Example of a "glitch" on a logic line. Glitches can cause
         fast TTL systems to misbehave, and their avoidance requires
         careful and neat layout of circuit boards and adequate
         decoupling of power lines close to each TTL device.

The speed with which a TTL gate can operate is most usefully
measured by the gate propagation time, which is the time taken for a
change in one input level to bring about a change in an output level
(the level change itself occurs much more radpidly). For the 74 family
the gate propagation time is approximately 10 ns, so that circuits
built up from 74 family devices can handle signal levels changing at
regular rates up to about 35 MHz, although the normal design maximum
is best restricted to 25 MHz. This is a usefully high speed and,
together with the simplicity with which devices may be connected to
one another, accounted for the early popularity of the 74 series. The
price paid for this high switching speed is the power consumed by the
gates, which is about 10 mW per gate. This doesn't sound like a lot of
power, but it must be remembered that each package can house several
gates, and that complex systems may use hundreds or thousands of
gates, so that the power drawn by a large system based on the 74
family can become significant. In practice 74 family packages operate
warm to the touch, and when a large number of packages are housed in a
small instrument case the resulting heat generation can cause problems
for other devices, such as op-amps.

In practice the very high switching speeds of TTL gates and the
large current spikes that gate operation generates on the power supply
lines (one of which is ground), can be the cause of some difficulty.
One of the secrets of success with TTL is to keep power line and
signal line inductances as small as possible to minimise the
generation of glitches. (Glitches are extremely fast pulses produced
in one gate during the operation of another gate. It is sometimes
difficult to see these glitches on an oscilloscope unless the 'scope
is both fast and provides time-base control that allows you to examine
both ends of a main pulse. A typical glitch is shown in fig 4.5, and
while they may be difficult to see, TTL gates manage to respond to
them all right!). Unfortunately keeping inductances low is
incompatible with bread-boarding a circuit with lots of
interconnecting wires. As a result the development and testing of TTL
designs is not always a joy, and it usually causes far less
aggravation to build a high speed system with a carefully planned
printed circuit board in the first place.

Table 4.3   TTL families and principal characteristics

family name	typical code	max switching speed /MHz	power per gate /mW
Standard	7400	35	10
High power	74H00	50	22
Low power	74L00	3	1
Schottky	74S00	125	19
Low power Schottky	74LS00	45	2
Advanced low power Schottky	74ALS00	90	1

Over the years other families of logic gates have been developed from the basic TTL designs. A high power family was produced to allow greater switching speeds at the expense of greater power dissipation, and a low power family offered lower power dissipation although operating at lower switching speeds. The addition of Schottky diodes across most of the transistors which make up normal TTL gates results in a fairly dramatic increase in speed - although again at the expense of higher power dissipation - and provides the Schottky TTL family. Increasing the impedance levels throughout the structure of each gate slows the devices down but enables them to operate with lower currents. This technique coupled with the use of Schottky diodes has resulted in the development of the Low-power Schottky family, which is only slightly faster than the original 74 family but offers the advantage of much lower power requirements (about 1/5th of that of the 74 family). In the late 1970s an Advanced Low-power Schottky family was introduced, offering double the speed and half the power demand of its predecessor, and this would probably be the best family of gates to adopt if one is going to hold stocks of the devices. Of course the newer families do tend to be more expensive than the older ones, but the cost of most SSI ICs is trivial compared with the cost of making something useful with them.

Each family of devices can be identified by its sensible identification code. Examples of these codes and the speed and power characteristics of each of the TTL families are summarised in table 4.3. All the families have a fanout of ten with gates of the same family, but care is needed if gates of different families are to be connected together because the lower power families are capable of sinking less current than the others. Data sheets should be carefully

studied if members of different families are to be connected with
fanouts of greater than two. Manufacturers usually recommend that the
inputs of any unused gates should be connected to the +5 V supply, to
prevent spurious level changes of the unused output from contributing
spikes to the power rails. On no account should unused inputs be
grounded - this only causes the package to use more power than is
necessary. Devices which belong to the 74 families are specified for
operation over the (package) temperature range of 0 - 70 C. The same
gates specified for operation over the military temperature range (-5!
to +125 C) are identified as the 54 families.

4.3 CMOS families

The other class of logic gate families of major importance is
that based on CMOS technology (Complementary Metal Oxide-Silicon). The
modern families available in this class are all low-cost, with low-
power requirements and are considerably more pleasant to use than the
TTL families. CMOS gates have very high input impedances in both hi
and lo states (they use fet-type inputs), so that virtually no
currents pass between gates and the power supply spikes characteristic
of TTL level changes are virtually non-existent. As a result the
chances of a circuit misbehaving through pick-up of spurious signals
from other gate connections or power rails are very much smaller than
with TTL. Some of the CMOS families have been designed as a pin-for-
pin replacement for the TTL families. One is identified with the
coding 74C, and the latest arrival is the high speed family identified
as 74HC. As this latter has the speed of 74LS TTL together with the
advantages of CMOS it is undoubtedly destined to become very popular.
However, the most widely used CMOS families are called the 4000
families and we shall concentrate on one of these, noting in passing
that the 74C and 74HC families share most of its characteristics.

The original 4000 family was developed by RCA, but soon replaced
by a modified form called the A-series, with its packages designated
4001A etc. This series had a few minor problems, such as gates which
did not work, and packages which burnt out if the power supplies were
turned on in the wrong order! Most of the A-series has now been
replaced by the B-series (the B is for "buffered" apparently) and
these are the CMOS family of choice for most systems. Some
applications (such as level changing) call for unbuffered gates and
the improved versions of these, which otherwise meet the B
specifications, are identified as UB devices. Examples of some of the
B-series gates are given in table 4.4.

The 4000B series gates use very little power and most will
operate with power supply levels anywhere within the range of +3 to
+15 V. Thus they can be used easily in battery powered instruments,
although it must always be borne in mind that if a gate is used to
supply current to something, such as a resistive load or a TTL family

Table 4.4   Examples of CMOS 4000B series devices

code number	no. of devices per package	type of device
4011B	4	2-input NAND
4023B	3	3-input NAND
4012B	2	4-input NAND
4068B	1	8-input NAND
4001B	4	2-input NOR
4081B	4	2-input AND
4071B	4	2-input OR
4070B	4	2-input EOR
4049UB	6	inverters
4050B	6	buffers
4016B	4	switch
4040B	1	12-bit counter
4060B	1	14-bit counter
4515B	1	4-bit decoder

logic gate, then the power to drive that load comes from the gate's power supply, and can easily be hundreds of times greater than the power needed to operate the CMOS gates alone. The devices are capable of only low current outputs, typically sourcing up to 0.2 mA in the hi state and sinking up to 0.5 mA in the lo state when operating with 5 V supplies, and about three times these values at 10 V. (Higher currents are avaiable from the 4049 and 4050B devices, see section 4.4.) Furthermore the low gate power advantage is only realisable when all the gate inputs are connected to either a hi or a low level. Unused inputs on a package must not be allowed to float – a rule which is even more important for CMOS gates than for TTL. The best course is to connect any unused inputs to the positive supply line.

One consequence of the lower gate power demand is that the CMOS gates have slower switching speeds than the faster of the TTL families (although the 74HC family seems to have overcome this problem), and the switching speed is dependent on the power supply voltage. For the SSI devices the maximum switching rate should be limited to 5 MHz with a 5 V supply and 10 MHz with a 10 V supply. Generally it is inconvenient to use anything other than a 5 V supply in a circuit which is to be connected to another system (such as a microcomputer), because of the necessity of ensuring that the levels used by the two systems are compatible. In spite of this, the advantages in terms of higher speed and greater output current capability are often

sufficient to justify operation at 10 or 12 V, even if an extra
circuit is required to change levels for interfacing to TTL.

The hi and lo logic levels of the CMOS 4000B series depend on the
power supply applied to the packages. The hi level is nominally equal
to the positive supply voltage, and the lo level is nominally ground.
To within about 1% these are the levels which the gates use for
output, and the high input impedance of CMOS gates means that these
levels are also those that will exist on gate interconnections – so
there is no level degradation produced by gates loading one another.
However, the input of a CMOS gate recognises the level change at
almost precisely one half of the supply voltage; this is a
considerable improvement over the TTL families and bestows upon the
4000B series a much better immunity to interference. Furthermore the
switching time of a CMOS gate is longer than the gate propagation time
(ie. the time between input and output level changes) and this tends
to result in CMOS systems absorbing glitches and system noise – a
particularly welcome feature after the horrors of some TTL circuits.

4.4 CMOS and TTL together

For some reason one nearly always seems to need a TTL device in a
circuit otherwise built exclusively with CMOS devices – or vice versa.
Fortunately interconnection is possible provided that the different
impedances and current capabilities of the two classes are taken into
account. If both devices are operating with a +5 V supply then CMOS
output will directly drive a single 74LS input or two 74L or 74ALS
inputs (as shown in fig 4.6a). If standard TTL needs to be driven then
one of the special TTL-driver CMOS buffers must be used, as
illustrated in fig 4.6b. An output from any of the TTL families can
drive dozens of CMOS inputs, although the TTL output must also be
connected by a 2-10 k pull up resistor to the +5 V supply line to
ensure an adequate hi level value (fig 4.6c). The special TTL-drivers
are required to interface TTL with CMOS circuits that are operating at
supply levels greater than 5 V, as in fig 4.6d. The 4050B (buffers)
and 4049UB (inverters) are somewhat unusual in that they allow an
input signal level to be greater than the supply voltage, although the
output hi level is determined by the supply voltage. The 7406
(inverters) and 7407 (buffers) have "open collector" outputs,
requiring a resistance pathway to the positive supply voltage. This
allows the hi output level to approximately equal that supply voltage,
and these devices can be used to interface TTL with CMOS operating at
>5 V as illustrated in fig 4.6e.

The 74HC CMOS series deserves a special mention because of its
remarkably high speed capability (50 MHz). Unlike the other CMOS
families the 74HC family is limited to a supply voltage of 5 V (7 V
absolute maximum). However, it is capable of driving a standard TTL
input or up to 10 74LS inputs, and may itself be driven directly by

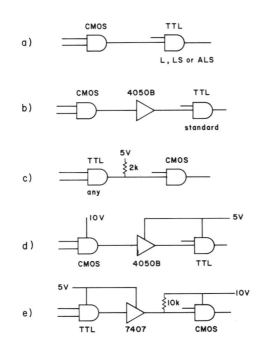

Fig 4.6   Interfacing TTL and CMOS devices. a) CMOS (at 5 V) will drive
the low power TTL families directly, but b) a high power
buffer is required to drive standard TTL. c) TTL will drive
lots of (5 V) CMOS inputs with a pull-up resistor. When CMOS
is operated at 10 V, high voltage buffers must be used to
interface with TTL, d) and e).

TTL levels without pull up resistors (its logic input levels having
been designed to be TTL compatible).

4.5 MSI circuits

Both the TTL and CMOS systems have been developed well beyond the
level of complexity required for the SSI gates introduced above. A
wide range of ICs is now available using this medium scale integration
(MSI) technology, and these make possible the manipulation of the 0
and 1 signal levels in almost any desirable manner. Discussing even a
representative selection of the available MSI circuits is beyond the
scope of this text, so that we shall make use of just two examples of
this type of device. However, most manufacturers publish data books
which provide details of all the ICs offered in their TTL and CMOS
ranges, and there are several books which discuss examples of the most
popular devices in a thoroughly digestible manner (see bibliography).

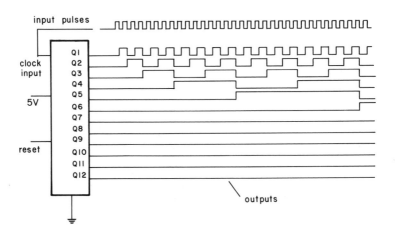

Fig 4.7 The 4040B binary ripple counter, showing how the outputs (Q1
         - Q12) change level in relation to a sequence of pulses
         applied to the input.

4.5.1 A 12 bit counter

     The first device that we shall examine has been chosen because o
its particular relevance to laboratory applications of microcomputers
its function in that connection being discussed in chapter 6. The
device is a binary ripple counter, a type of device available in a
variety of forms both as TTL and CMOS circuits. The particular form we
shall discuss is the 12-stage binary ripple counter of the 4000B-
series, the 4040B. A schematic representation of the device is shown
in fig 4.7. The device has an input (called the clock input, for
reasons which will become apparent) which accepts CMOS 0/1 levels, and
twelve outputs (designated Q1 - Q12) which produce 0/1 levels. The
connection called "reset input" is normally held at 0, ie. grounded.
The levels of the twelve outputs can be regarded as representing a
binary number of twelve BInary digiTS (bits). The pattern of twelve 0
and 1 levels (such as 001000101110, in which, by convention, the
highest order bit - Q12 - is written first, just as for normal decima
numbers) forms a 12 bit binary number, which can have a decimal
equivalent value in the range 0 (000000000000) to 4095 (111111111111)
Thus ouput Q1 can be 0 or 1 and can contribute 0 (when Q1 is 0) or 1
(when Q1 is 1) to the value of the binary number. Output Q2 can
contribute 0 (when Q2 is a 0) or 2 (when Q2 is a 1), output Q3 can
contribute 0 or 4, and so on up to Q12 which can contribute 0 or 2048
(2^{11}). The number represented by the outputs (the "count") is
incremented by one each time the clock input changes from a 1 to a 0
level. Thus the device counts the number of such "negative
transitions" on the clock input, and as there is one negative

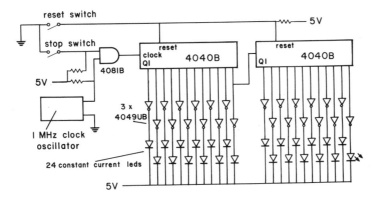

Fig 4.8   A fast timer, using a 4040B to count the logic pulses from a
          1.0 MHz crystal oscillator.

transition each time the clock input goes through the cycle 0-1-0 (ie.
each time a positive-going logic level "pulse" arrives at the input)
the device can be used for counting pulses. The count can be reset to
zero at any time by the application of a 1 level at the reset input,
and counting starts again as soon as the reset input is returned to
0.

An alternative way of regarding the 4040B is as a divider, in
fact it is often referred to as a divide-by-4096 counter. The output
Q1 changes only when the input changes from 1 to 0 (and not when the
input changes from 0 to 1), so that when the input is "clocked", ie
when a continuous stream of alternating 0s and 1s is applied to the
input as illustrated in fig 4.7, the output Q1 changes half as
frequently as the input - so it divides the input clock frequency by
2. The output Q2 divides the clock frequency by 2^2, ie. 4, and the
output Qn divides the clock frequency by 2^n. The highest order output
is Q12, which divides the output by 4096 (2^{12}).

We can illustrate these two views of the 4040B by outlining two
circuits which utilise these principles. The first, shown in fig 4.8,
is a fast timer which counts the pulses generated by a 1.000 MHz
crystal clock oscillator. Such oscillators produce highly accurate
TTL-compatible squares waves, and are available for a wide range of
frequencies from kHz to MHz. The pulses are gated into the counter's
clock input using an AND gate as a start/stop switch, and the 12 bit
binary output is displayed using 12 LEDs. (The LEDs are buffered using
the 12 4049UB high current inverters. Note that the inverters are used
so that the LEDs light on a lo level output. In this state the 4049UB
can sink 5 mA when operating with 5 V supplies. TTL buffers could be
added is higher currents were required.) The Q12 output is also

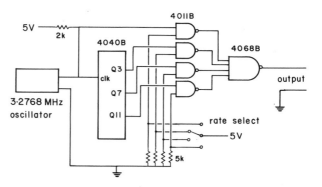

Fig 4.9   Using a 4040B to divide a 3.2768 MHz pulse rate by a selecte
          amount.

connected to the clock input of a second 4040B, whose 12 outputs are
displayed on a second dozen LEDs. Thus the two counters form a 24 bit
binary counting system which can register the time between start and
stop in the range 1 to 2^{23} (about 16 million) microseconds. A simple
manual reset switch is provided to reset both counters to zero. In
this illustration the start/stop switch is manually operated, althoug
it would be more usefully activated by a signal level derived from a
laboratory apparatus, and the binary output is just a pattern of
lights, although it would be an ideal candidate for transmission to a
microcomputer. However, its the 4040B's role which interests us at
present - we shall return to these other matters in due course.

     It should be noted that the outputs of this circuit provide an
example of a parallel digital signal - parallel because the 0/1 level
are carried on different conductors, side-by-side, but at the same
time, and digital because the pattern of outputs is used to represent
a number.

     The second 4040B application example, illustrated in fig 4.9, is
a pulse generator for producing 0-5 V oscillations at selectable
frequencies of 6.4, 51.2, 409.6 and 3276.8 kHz. (The extension to a
larger number of frequencies using the other outputs should be
obvious.) In this example the signal from a 3.2768 MHz oscillator,
which also provides the 4040B's input, and the output signals from Q3
Q7 and Q11 are fed to four 2-input NAND gates, the other input of eac
being set hi or lo by a "rate select" rotary switch. Only one of the
NAND gates can have both inputs hi, and therefore only one NAND gate
can have its output go lo. The outputs from the four NAND gates are
combined using a 4-input NAND gate. This gate always has three inputs
hi and one input changing at the selected frequency, so its output
also changes at the selected frequency, being lo when all its inputs
are hi and hi when one of its inputs is lo.

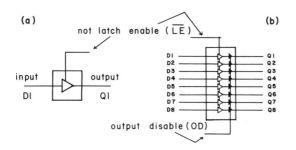

Fig 4.10 a) a single latch. b) the 74C373 octal latch with output
disable facility.

Circuits of this kind are valuable for providing pulses of highly
reproducible duration, varying on/off times over a very wide range, or
dividing down a pulse rate before it is measured on a ratemeter
(while, incidentally, maintaining the statistical precision of the
undivided pulse rate - see chapter 2 on shot noise).

4.5.2 The TRI-STATE® latch

The second device we shall discuss for handling digital signals
is the TRI-STATE® latch, a device of particular importance for the
transmission of digital signals to or from a microcomputer. While we
shall discuss one particular device, the 74C373, the following points
should be noted:

1) Most of the principles of latching and TRI-STATE® output apply
   to a wide range of other devices designed to generate parallel
   digital outputs.
2) TTL versions of this device are available but are not suitable
   for some of the interfacing applications discussed in chapter
   6.

Consider first an 8 bit parallel digital signal on 8 conductors,
which we shall refer to as lines Dl - D8. Imagine that the signals on
these lines are changing rapidly, as they would, for example, in the
case of eight of the outputs from the timer circuit of fig 4.8. If we
wished to examine the signals on the lines (eg. by looking at the
display LEDs of fig 4.8) we would need to stop the signals from
changing while the examination was in progress. Of course, stopping
the timer of fig 4.8 also terminates our ability to continue with the
time measurement, so that stopping the signals from changing by this
method could be undesirable. An alternative technique is to pass the
signals into a group of special buffers known as latches.

A single latch, illustrated in fig 4.10a, has an output, Q1, whose level follows the input, D1, as long as the level of the "not latch enable" pin is hi. When the level at the "not latch enable" pin is taken lo, then the output Q1 remains at the same level that it had when the "not latch enable" pin was hi. In other words the latch remembers the value of the input level at the moment "not latch enable" dropped from hi to lo. Now the words "not latch enable" tend to become a little cumbersome after a while, and the name latch enable is more commonly used to identify the pin. However, changing the name can cause confusion over whether latching occurs when the signal at that pin changes from a 0 to a 1 or vice versa. When the pin is called, correctly, "not latch enable", a hi or "true" level indicates that the latch is not enabled - ie the lock on the door between input and output has not been applied, so signals can still pass through. A lo or "false" level at that pin means that the latch has operated and data can no longer enter through the inputs.

The 74C373 is a CMOS version of an octal latch and so has 8 latches in its 20 pin package, as illustrated in fig 4.10b. The 8 inputs (D1 - D8) feed 8 outputs (Q1 - Q8), and all 8 may be latched a the same moment by a hi to lo transition on the latch enable pin. Thu the value of an 8 bit number can be latched and examined on the outputs, Q1 - Q8, while the signals on D1 - D8 continue changing, allowing, for example, the timer of fig 4.8 to continue operating. When the latch enable pin returns to a hi level the outputs of the device will once again follow the inputs until another latch enable signal is applied.

Eight conductors can be squeezed into a fairly small space these days, but it would nevertheless be inconvenient to rely on a large number of multiwire cables for the transmission of parallel digital signals between items of laboratory instrumentation. Fortunately ther is a remarkably simple alternative known as a "bus". A bus is a collection of conductors for parallel signals which is connected to several different sets of inputs and outputs, so that any of the inpu sets can receive signals from the bus, and any one of the output sets can transmit signals to the bus. For this system to work it is, of course, important that only one output at a time determines the value of the signal level on a single conductor. The signal levels we have met so far are not suitable for use in bus operation because there ar only two allowed levels (hi and lo), and if one output on the bus is trying to force its attached conductor to lo, while a number of other outputs on the bus are trying to force the same conductor hi, then a lot of power is being expended with no certainty as to who will win.

One solution to this problem is to ensure that all outputs connected to the bus are open collector types. In this case each of the bus lines will be hi unless one of the device outputs is assertin a lo level on the line. Of course, it is up to the circuit designer t

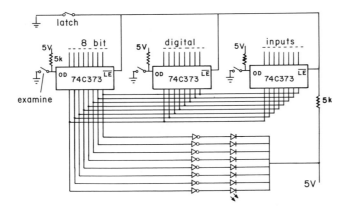

Fig 4.11 Connection of the outputs of several 8-bit TRI-STATE® devices
to a common bus - in this case feeding 8 buffered leds
through high current inverters.

ensure that only one device is attempting to assert lo levels at any
one time. The disadvantage is that CMOS gates do not offer a open
collector output variant, so that one of the TTL families must be
used.

An alternative and elegant way around this difficulty, developed
by National Semiconductor Corporation, is to introduce a third state
in which an output can happily exist while some other output
determines the level of the attached conductor. This third state is
variously termed the "off" or the "high impedance" state. The
important point is that in this third state the output does not effect
the signal level on the conductor attached to it. Although this
technique is called TRI-STATE® logic, it is important to remember that
the signal levels remain the same as conventional TTL or CMOS levels
(depending on the device); it is a third output state which has been
added (an off state) and not a third signal level.

In order that the outputs may be directed into the high impedance
state, TRI-STATE® devices have a pin connection called "output
disable". With a lo level applied to the output disable pin, the
devices outputs behave just like any other logic outputs. When a hi
level is applied to this pin all the devices outputs are switched into
the high impedance state. Of course, if this facility is not required
at all then the output disable pin can be permanently grounded. The
74C373 has TRI-STATE® outputs and these are compatible with CMOS and
low-power TTL inputs (with a fan-out for the latter of 1).

Figure 4.11 shows three octal latches with their inputs derived

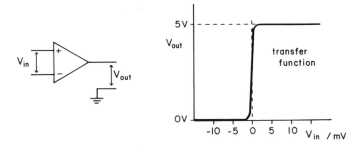

Fig 4.12 Circuit symbol and transfer function of a voltage
comparator.

from a suitable 24 bit parallel digital signal source, such as the
outputs of fig 4.8. The outputs of the latches are connected into a
common bus which also connects to eight buffered LEDs. The latch
enables are connected together and operated by a single switch which
can be manually operated to latch the data on the 24 input lines. The
three output disables are connected to separate "examine" push
switches, so that whichever switch is operated the corresponding 8
bits of latched data are placed on the bus and displayed on the LEDs.
Study this circuit until you fully understand its operation. There are
more of these to come.

4.6 Generating logic levels

We have already met the TTL compatible crystal oscillator
packages suitable for generating 0/5 V pulses. These and 5 V dc
supplies provide the main source of logic signal levels. However, a
major requirement in the application of digital and logic circuits is
often the generation of logic level signals from some other kind of
signal. For example, many types of transducers generate pulses which
are counted when the transducer is used for measurement purposes.
However, the pulses generated by the transducer are rarely precise
changes from 0 to 1 levels and back, so that some kind of interface
circuit is normally required to translate the non-logic signal levels
to the required logic values. Several types of device are available
for this interfacing role, but one of the most widely used is the
comparator.

4.6.1 The comparator

A comparator is superficially similar to an operational amplifie
- in fact op-amps can be used as comparators in some cases. However,
comparator is designed to produce an output logic level dependent on
the differential input at its inverting and non-inverting inputs.

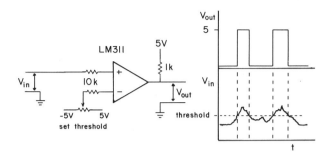

Fig 4.13 The use of a comparator as a threshold detector. The circuit
         produces a logic 1 output while V_{in} is greater than the
         threshold level. (All resistors 2 k.)

(Generally the output is TTL compatible, and should also serve CMOS
operating with 5 V supplies.) An example is shown in fig 4.12, where
the transfer function (ie. a curve showing the output as a function of
the differential input) illustrates that the output is a logic level 0
whenever the non-inverting input is more than a millivolt lower in
potential than the inverting input, and a logic level 1 whenever the
non-inverting input is at the higher potential. As the comparator's
input signal is a differential signal, a comparator can be used for
generating logic level signals in response to the variation of an
analog input signal compared with some preset voltage level. A simple
example of this use is shown in fig 4.13, where an analog input
signal, V_{in}, is applied to the non-inverting input and is compared
with the "threshold" level applied to the inverting input. Note that
the 311 has an open collector output, in this case taken through a 2 k
resistor to a TTL hi level, so providing an output compatible with
both TTL and 5 V CMOS. By changing the hi reference and resistor value
(to, say, 10 V and 10 k) this device could also drive CMOS operating
with 10 V supplies. Not all comparators provide open collector
outputs.

     Comparators suffer from many of the imperfections discussed for
operational amplifiers, and generally speaking require relatively high
input bias currents. However, comparators are designed to be used
without feedback loops (although some feedback can be useful to
introduce an element of hysteresis into the transfer function), and to
produce an output which can change level fairly rapidly through the
range 0 to 5 V (most comparators cannot generate a negative output
voltage, even though many require a negative supply voltage). The
imperfection which is peculiar to comparators is a variation in the
rate at which the output can change as a function of the differential
input voltage - a variation which is often different for positive and
negative input steps. This imperfection is characterised by

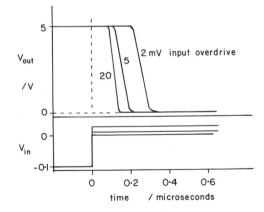

Fig 4.14 The variation of the response time of a typical voltage
comparator for various input overdrives.

manufacturers in diagrams showing the response time (which is actual
the time delay between an input change and an output change) for
various input overdrives (the amount by which the differential input
voltage exceeds zero). A typical selection of such curves are shown
fig 4.14, illustrating the kind of data which needs to be considered
before a particular comparator is chosen for a circuit. Clearly a sl
device with the characteristics showns in fig 4.14 would not be a go
choice for use in a circuit which would need to produce pulses at a
rate close to 10 MHz. The basic properties of a number of comparator
are collected in table 4.5.

Figure 4.15 shows a circuit for a light detector designed to
produce a logic pulse when the photomultiplier senses a scintillatio

Table 4.5   Characteristics of some typical voltage
comparators

Device no	306	311	319	339	360
Devices per package	1	1	2	4	1
bias current /microamps	25	0.25	1	0.25	15
response time /ns (100 mV overdrive)	40	200	80	1300	16
TTL fanout	10	5	2	1	2
CMOS >5 V ?	no	yes	yes	yes	no

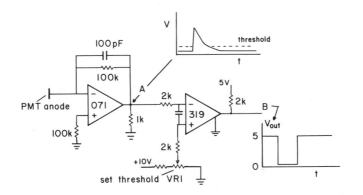

Fig 4.15 The production of logic pulses from a photomultiplier tube
monitoring scintillations. A transresistance amplifier is
connected to a comparator with a threshold level set by VR1.
The threshold may be set to determine the minimum
scintillation intensity which gives rise to an output pulse.

(a brief flash of, in this case, about twenty photons). The PM tube is
being operated with a (virtually) grounded anode, and the pulses of
negative charge arriving at the inverting input of the transresistance
amplifier produce positive going voltage pulses at its ouput, A. These
signals are passed through an input resistor to the inverting input of
a fast comparator, whose non-inverting input is held at a suitable
threshold voltage by VR1. Under dark conditions the voltage at the
comparator's inverting input is lower than the threshold voltage, so
the comparator's output, B, is at a TTL 1 level. When a scintillation
produces a voltage pulse at A which exceeds the threshold voltage, the
comparator's output changes to a 0 and remains there until the voltage
at A has decayed to a value below the threshold. Thus the signal at B
consists of negative-going pulses (ie. the level goes from 1 to 0 when
a pulse is initiated; the voltage level never goes negative). The
width of each pulse depends on how long the voltage at A remained
above threshold, so the pulse widths are not constant, although this
does not matter if the pulses are only to be counted. Indeed
measurement of the pulse widths may be used to determine the
intensities of detected scintillations.

The analog signal at A is useful if the signal needs to be
examined on an oscilloscope, or if pulse height analysis is to be
performed. When connections of this kind are to be made to circuits
handling ac or pulse signals is important to check that the connection
does not effect the operation of the rest of the circuit or, worse,
grossly distort the analog output, by the introduction of additional
capacitative impedance. With the component values shown the circuit
counts scintillations at up to about 10^4 per second with dead-time

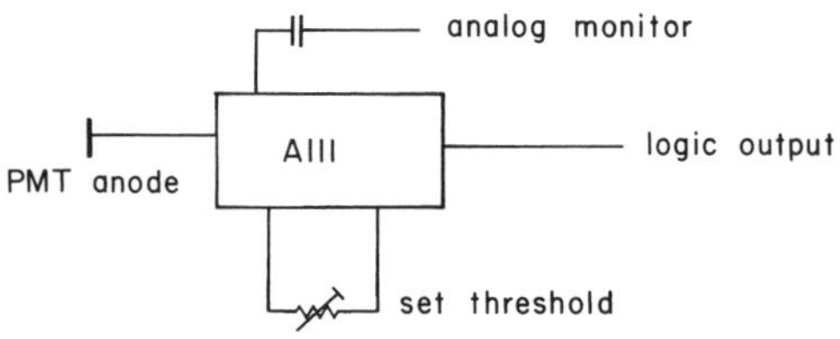

Fig 4.16 The simple arrangement possible when using a charge amplifie
        discriminator, such as the AmpTek All1, for the detection of
        photomultiplier pulses.

losses of <3%. Faster devices could be used to improve this figure,
although to achieve low dead times at pulse rates above 1 MHz require
the use of narrow (<100 ns) pulses with even shorter (<20 ns) rise
times. Great care needs to be taken with the layout of circuits
intended to handle pulses in this frequency region.

A considerable improvement in speed over the circuit of fig 4.15
can be obtained by the use of a single IC device designed specificall
for the conversion of low-level charge pulses to logic pulses. A
circuit using one such device is shown in fig 4.16, where the device
is the Amp-Tek PAD (pulse amplifier discriminator) All1. The circuit
does the same job as that described above - except that it counts up
to 10^5 Hz without significant dead time. The device is excellent, and
expensive, and very delicate - touching its pins with a test signal
can destroy it.

4.6.2 The monostable, astable and bistable

Another valuable device for the production of logic pulses is th
monostable, a device which prefers to retain its output at one level
although it can be forced into the other level for a preset period of
time. Most of these devices are useable in two slightly different
ways, depending on connections made to the device, and the devices ar
referred to using the broader name "multivibrator". Firstly, they can
be used to produced a single logic pulse of specified width in
response to some input trigger. An example is shown in fig 4.17a,
where a 4047B (CMOS) multivibrator is being used as a monostable,
producing output pulses in response to positive going transitions at
the +trigger input. (A -trigger input is also provided so that the
device can respond to negative going transitions.) Both positive goin
and negative going output pulses are available from the device's
complementary outputs, and the width of the output pulse is 2.48 CR s
The monostable finds application in reducing a stream of variable
width pulses to a series with uniform width, and in producing logic
pulses from non-logic type signals - although this particular device
is triggered by "edges" of voltage change, which have to be quite fas
(typically <10 microseconds).

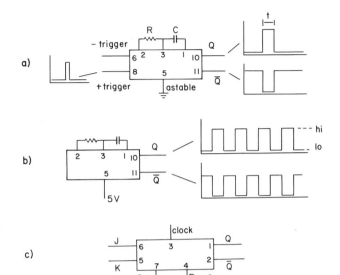

Fig 4.17 a) Monostable operation of the 4047B, producing a single
pulse each time the device is triggered. b) Astable
operation, generating a continuous stream of pulses as a
square wave. c) The 4027B bistable JK flip-flop device.

The second way in which this device may be used is as an astable
multivibrator and in this guise it produces a continuous stream of
pulses. The 4047B is capable of astable operation only with a 50% duty
cycle (ie. the times spent by the output in the hi and lo levels are
equal), so the output appears as a square wave. An example of this
mode of operation is given by fig 4.17b, where the frequency of the
output pulse stream is governed by CR and the period is equal to 4.4
CR s. The 4047 can be used to provide pulse streams in the frequency
range 1 kHz to 1 MHz and is therefore a useful device for the
production of clocking signals, although the frequency stability is
typically only a few percent – which is several orders of magnitude
poorer than can be achieved with a crystal oscillator.

A closely related device is the bistable or flip-flop circuit.
There are in fact several varieties of these, although they share the
properties that their outputs can remain hi or lo (ie, they have two
stable output states), and that their outputs can change levels only
at particular times in relation to a clock signal appiled to the
device. It is this latter property which defines the flip-flop as the
basic circuit of a system known as clocked- or sequential logic, and
the fundamental building block of the bulk of MSI circuits, such as
the binary ripple counter we have already discussed.

Table 4.6   Truth table for the 4027B JK flip-flop

inputs J	K	clock	output Q
0	0	0-to-1	Q (no change)
1	0	0-to-1	1
0	1	0-to-1	0
1	1	0-to-1	Q (toggle)

A typical flip-flop is illustrated in fig 4.17c. This one has two inputs in addition to its clock signal input, and these are called the J and K inputs (although there does not appear to be any particular reason for this nomenclature) and the device a JK flip-flop. The important properties of a JK flip-flop are:

1. If the outputs are to change they can only do so when the clock signal goes from lo to hi.
2. The outputs are determined by the J and K inputs according to the rules summarised in table 4.6.

In addition JK flip-flops such as the CMOS 4027B have two other inputs which force the outputs into particular states irrespective of the clocking signal. Thus a 1 on the SET input forces output Q to a (and the complementary output to 0), and a 1 on the RESET input forces Q to a 0.

This type of device has a wide variety of applications in addition to forming the basis of more complex devices. For a start it is the basic "divide-by-2" unit, in that with J and K both hi, the output changes once for every two level changes (ie. hi-lo-hi cycle) on the clock input. (We use this property in fig 6.12) Furthermore the device can be used to synchronise one level change with another, a particular valuable feature in timing applications. For example, if J=0, K=0 and Q=0, then a change to J=1 will not result in a change to Q=1 until the clock signal changes from lo to hi. Similarly if J=0, K=1 and Q is forced hi by a brief SET=1, then the output, Q, returns to 0 only when the clock input changes to a 1. (We use this property in fig. 6.16.)

4.7 Analog/digital interconversion

Digital techniques offer ways of handling and transmitting signals rapidly and reliably, and are highly immune to the effects of interference. Microcomputers are exclusively digital, but the

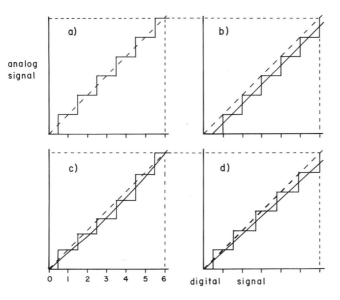

Fig 4.18 The mapping between an analog and a three bit digital signal,
showing a) the inevitable quantisation error, b) a zero
error, c) a non linearity error, and d) a full scale error.

laboratory world remains heavily dependent on analog signals. Most
transducers generate analog outputs, and many "results" are presented
in an analog form - such as spectra, chromatograms and other chart
records - even when quantitative data is digitally recorded. Clearly
the ability to interconvert analog and digital signals is of major
importance in the laboratory applications of microcomputers.

Performing these interconversions electronically requires
electronic circuits which map a dc signal (usually a voltage, say, 0 -
10 V) onto an n bit binary number between 0 and 2^n-1. This mapping is
illustrated in idealised form and on a highly enlarged scale in fig
4.18a, where the stepped function indicates the actual mapping. The
broken line drawn between 0 V, binary 0, and V Volts, binary 6,
represents the mapping which would result if fractional digital values
were allowed. The points at which the mapping is exact are those at
which the lines cross. At all other points the mapping is in error, by
up to the equivalent of half a digital unit, ie. +/- 1/2 LSB (a least
significant bit, 2^0). For an 8 bit binary number (0 - 255) this error
is about 0.2% of the maximum signal value, and for a 0-10 V analog
range is equivalent to +/- 20 mV on the analog scale. In general the
error is equivalent to 1 part in 2^n for an n bit conversion, and is
called the quantisation error of the conversion. Its effects are most
serious for small signals; for example, with a 10 V full scale range

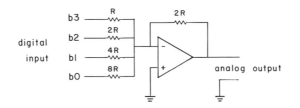

Fig 4.19 A typical digital to analog converter circuit.

an analog signal mapping to the digital value 2 may be 78 +/- 20 mV, an error of 25%.

This kind of conversion error is unavoidable when converting between a continuous analog signal and a discrete digital signal. Its effects may be reduced by operating in the range where the digital signal is a substantial fraction of the 2^n-1 maximum value, or increasing the resolution of the conversion (ie. the number of bits used for the digital signal). In practice the conversion error is mor serious than suggested by fig 4.18a because of imperfections in the circuits which bring about the conversion. These additional sources o error are many and complex, and a full description of them is beyond the scope of the present text. However, the effects of the three most important contributions to the additional conversion error, illustrated in fig 4.18b-d, are described briefly below - to sow a fe seeds of healthy scepticism about the accuracy of the conversion devices considered later.

Figure 4.18b shows the effects of a zero error, where the actual mapping is displaced from the ideal mapping. This error arises from input offset effects and may be easily corrected by offsetting the analog signal. In fig 4.18c we see the effect of a non-linear mapping function which can arise through circuit imperfections. At the point of maximum deviation from ideal mapping (which is -1/2 LSB in this example, the difference between the broken line and the solid curve) the converted signal may be in error by up to 1 LSB, although at the point of contact between the broken line and the stepped function the mapping is, of course, exact. So a specified non-linearity error of +/- 1/2 LSB implies a possible conversion error of +/- 1 LSB. Figure 4.18d illustrates the effect of a -1/2 LSB linear error (also called full scale error) in the slope of the mapping function. A full scale analog signal correctly maps to a digital signal of 6, but slightly smaller signals may have errors of up to 1 LSB.

4.7.1 Digital to analog conversion

There are several techniques for implementing digital to analog

Table 4.7   Characteristics of some popular DACs

Device no	type	resolution /bits	settling time /microseconds
ZN425E	R-2R	8	1
DAC0800	multiplying	8	0.1
AD7520	multiplying	10	0.5
AD7542	multiplying	12	2
DAC70/CSB		16	75

conversion. The principle of one of the simplest is shown in fig 4.19.
The circuit shown is for a 4 bit digital to analog converter (DAC)
accepting 0 or 5 V signals at its digital inputs, b0 - b3
(corresponding to binary inputs of 0 - 15), and producing an analog
output voltage in the range 0 - 10 V. The circuit consists of a
summing amplifier producing an analog output given by

$$V_{out} = 2b_3 + b_2 + b_1/2 + b_0/4 \text{ V}$$

There are two main difficulties with this approach. Firstly it
requires a number of resistors with very precise resistance ratios -
as these determine the linearity of the conversion. This becomes
difficult to achieve as the number of bits increases, and in practice
alternative techniques are used in the manufacture of single package
DACs. Secondly this technique relies on the analog voltage levels at
the hi digital inputs for the value of the analog output. This would
not be acceptable as, for example, the TTL logic levels can vary over
quite substantial ranges, and in practice the digital input hi levels
need to be converted to a precisely defined analog reference voltage
level.

The most popular single package DACs are those using a technique
known as the R-2R ladder (a variant of the fig 4.19 circuit). Most
devices working on this principle use an 8 or 10 bit digital input. An
alternative type of DAC is the monolithic DAC, which utilises bipolar
transistors to generate scaled currents from the digital signal
levels, and a transresistance amplifier to produce an output voltage
proportional to the sum of these currents. Monolithic DACs are
available as 8, 10 or 12 bit devices.

More sophisticated techniques allow up to 16 bit (or even 18 bit)
digital data to be converted, although at a significant cost. Some
DACs produce an output voltage which is proportional to the product of

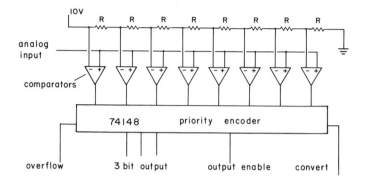

Fig 4.20 A typical analog to digital converter circuit based on the
parallel conversion technique.

an input reference voltage (or current) and the digital signal, with
the reference voltage variable over a wide range. Such "multiplying"
DACs allow the analog output range to be varied or scaled, and this
can be extremely useful where wide range analog output is required
with only limited resolution.

     As DACs tend to use analog output circuits there are limitations
on the rate at which the analog output voltage may change. Furthermor
the higher resolution DACs may require a significant time to carry ou
the conversion process and produce a stable output. Generally
conversion is initiated on the application of a logic level to a
"convert" pin, and the time required after this for the analog output
to become stable (ie. not to change by more than the equivalent of 1/
LSB) is called the settling time. The characteristics of some typical
DACs are collected in table 4.7.

4.7.2 Analog to digital conversion

     One of the many techniques used for converting an analog input
signal into a digital output signal is known as parallel conversion
(or flash conversion). This technique is used in some of the fastest
analog to digital converters (ADC) and an example of a converter base
on parallel conversion is shown in fig 4.20. The circuit illustrated
allows an analog input signal in the range 0 - 10 V to be converted
into a three bit binary output in the range 0-7, when the input enabl
(convert) line is taken lo. The chain of equal value resistors
provides eight equally spaced reference voltages between 1.2 and 10 V
inclusive, and the eight comparators produce outputs of 0 or 1,
depending on whether the voltage at their inputs is greater than or
less than the corresponding reference voltage. Thus an input signal o
5 V would produce 0s from comparators 1-4 and 1s from comparators 5-8

Table 4.8   Characteristics of some popular ADCs

Device no	type	resolution /bits	conversion time /microseconds
CA3300	parallel conv.	6	60 ns
TL507C	single slope	7	1000
ZN427E	succes. approx.	8	10
8703CJ	charge balance	8	1250
TDC1007J	parallel conv.	8	33 ns
AD7581	8 channel sa	8	20
AD574J	succes. approx.	12	25
ICL7109	charge balance	12	33 ms

This pattern of 0s and 1s is then encoded into a three bit binary number by a "priority encoder", which produces a binary output determined by the highest of the eight inputs to be at the 0 level, (irrespective of the signals on the lower number inputs). The outputs may be set equal to 7 at any time by taking the output enable line hi, (note that the 74148 does not have TRI-STATE outputs).

Parallel conversion ADCs are available with up to 8 bit binary outputs and are ideal when very fast conversions are required. Unfortunately they are expensive. The more humbly priced ADCs are slower devices and generally require an "initiate conversion" pulse on one pin, producing a "busy" signal on a second pin until the conversion is completed. One of the commonest ADC types uses a technique know as successive approximation. The analog input voltage is compared with an analog signal generated by a DAC from a digital value which the device adjusts, by successive approximation, until the two analog signals are within 1/2LSB. The digital value arrived at now forms the ADC output. This technique requires that the DAC changes its output value n times (for an n-bit converter), and so needs at least n DAC setting times to achieve conversion. To avoid the risk of the analog input signal changing during this time interval, successive approximation ADCs usually sample and hold the analog input signal at the start of the conversion cycle. Thus noise on the analog input (especially short duration spikes) can produce fluctuations on the digital output. Successive approximation ADCs are generally highly regarded for accuracy, although some can suffer from a problem known as "missing codes" - which is, essentially, producing the wrong answer once in a while.

The other main approaches to analog to digital conversion are based on the charging of a capacitor. This may involve using

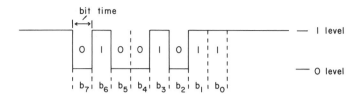

Fig 4.21 One example of serial digital data. In this case the data
        represents an 8 bit number.

standardised current pulses being counted into a capacitor until the
voltage across the capacitor exceeds the analog input voltage (single
slope integration). The count then forms the digital output.
Alternatively a charge balancing technique may be used. The most
popular variant of this involves using a current proportional to the
analog voltage to charge the capacitor for a fixed time, followed by
the discharging of the capacitor with a constant current and
digitising the time required to empty the capacitor (dual slope
integration). The problem with single slope integration is that the
conversion accuracy is limited by the accuracy and stability of the
capacitor and the comparator used. On the other hand it does provide
one of the best techniques for producing a uniform spacing between
adjacent portions of the conversion mapping. Dual slope integration
results in very accurate conversions which are not particularly
sensitive to the value or stability of the charge storage capacitor,
(the capacitor is usually "on chip"). The charge balancing techniques
are all fairly inexpensive, but slower than the successive
approximation method. Some typical ADCs are listed in table 4.8.

4.8 Serial digital signals

    Although we will not consider a specific serial data system unti
chapter 7, it is appropriate to consider the basics of serial digital
signals here. The parallel digital signals we have met so far have
consisted of signal levels on several different conductors – each one
representing a bit of the digital signal. As all the bits of a
parallel signal are available at the same time, time is not an
important element in determining the byte value of an 8 bit parallel
signal. Serial signals on the other hand consist of a series of signa.
levels applied sequentially to a single conductor, so that it becomes
essential to agree on the time span that each bit of the signal
occupies. A typical serial signal representing an 8 bit value is
illustrated in fig 4.21, where the duration of each bit is indicated.

    Communicating byte values bewteen devices in the form of serial
signals is attractive because it can be achieved with a simple twin
cable (signal conductor and ground). However, it does require

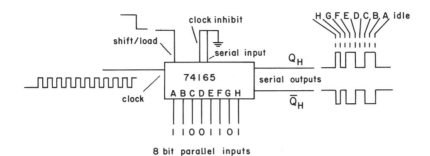

Fig 4.22 The 74165 8 bit parallel input/serial output shift register.

aggreement between the devices at either end of the cable on when each
transmission starts and on the duration of each bit. We will return to
these aspects of serial transmission in considering the RS232 data
communication standard in chapter 7. For now we confine our attention
to the generation of serial signals and serial/parallel
interconversion.

While in principle any sequence of level changes on a single
conductor may be regarded as a serial digital signal, for our purposes
the important serial signals are those used to represent 8 bit digital
values. The simplest way of producing a serial signal of this type
involves generating it from an 8 bit parallel signal, using an IC
device known as a shift register. A typical device, illustrated in fig
4.22, is the 74165, an 8 bit PISO (Parallel Input/Serial Output) shift
register. It has 8 parallel inputs (identified as A - H) through which
an 8 bit parallel digital signal may be latched into the register by a
hi-to-lo transition on the shift/load pin.

When the shift/load pin goes lo the serial output, Q_H, goes to
the level stored in the register's H bit (and the complementary output
is at the other level). When shift/load goes hi the register is able
to shift its bits one bit to the right on the rising edge of each of
the pulses applied to its clock input. The old content of H is
discarded and H receives the old G bit, G receives the old F, F
receives the old E and so on. A receives whatever level is present on
the serial input pin, so if this is permanently grounded A becomes 0.
As the output Q_H produces the new value of H, this output shifts
through the sequence of levels representing the bits H down to A
during 8 clock cycles. In our example the serial input pin is
grounded, so once the 8 bits of data have been shifted out of the
register Q_H remains at 0 (and the complementary output at 1).

The reverse translation process is achieved using a Serial
Input/Parallel Output (SIPO) shift register, such as the 74164

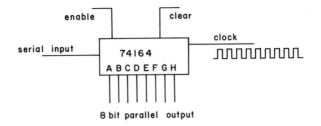

Fig 4.23 The 74 164 8 bit serial input/parallel output shift register.

illustrated in fig 4.23. This device has 8 outputs which are
determined by the corresopnding bit values in its register. A lo
signal applied to the clear pin sets all bits to 0. The device has two
serial inputs. When both inputs are hi then a 1 is placed into the
register's A bit on the rising edge of the next clock pulse when the
register shifts right and the old content of A is moved to B, the old
content of B is moved to C and so on. The old H is discarded. When
either of the two serial inputs is lo a 0 is forced into the A bit on
the rising edge of the clock signal. This arrangement makes it
possible to use one of the serial inputs to enable the other (as if
either one is held lo the 8 parallel outputs become 0s after 8 clock
cycles). Of course, the desired 8 bit parallel output only has its 8
bits all present during the eighth clock cycle, so its up to the user
to ensure that this byte is transferred to a latch at the right time.

CHAPTER 5

THE MODERN MICROCOMPUTER

Digital signal handling circuits continued to develop from the
MSI devices described in the last chapter until it became possible to
produce thousands of gates in a single device - a technology known as
large scale integration (LSI). During the late 1970s a number of
companies began using these devices to produce small desk-top machines
which had many of the capabilities of the available computers: the
microcomputer age had dawned. Since that time a wide variety of
microcomputers has become available. Some are designed for the
business market, for use in word-processing, record keeping and
accounting. Others are aimed primarily at the personal or hobby market
- and often that means game playing and some useful software for
graphic displays. With the exception of some of the Hewlett Packard
range, few are actually intended to be used for instrumental purposes
in scientific laboratories, although many can fulfill this role
surprisingly well.

5.1 The eight bit micro

Most of the microcomputers currently used in laboratory
applications are 8 bit computers, i.e. the numbers stored in the
computer's memory and used for communication with external
instrumentation are handled as "bytes" of 8 bits (BInary digiTS) each.
A byte representing a binary number is shown in table 5.1. The byte is
stored within the computer as eight digital signals, each of which can
be either a 1, taken to represent a binary 1, or a 0 representing a
binary 0. When shown diagramatically as in table 5.1 the eight bits
are referred to as bit 0 - bit 7, or b0 - b7. The binary value of the
byte is then just a list of eight 0's and 1's, and the equivalent
decimal value can be calculated by summing the contributions of each
bit, remembering that a 0 bit contributes decimal 0 while a 1 bit
contributes 2 raised to the power of the bit number. Thus a 1 for b0
contributes decimal 1 (2^0), a 1 for b2 contributes decimal 4 (2^2) and
a 1 for b7 contributes decimal 128 (2^7). The range of values which can
be held as a single byte is therefore 0-255. Some examples of 8-bit
bytes are shown in table 5.2.

Table 5.1 The 8-bit byte and its equivalent decimal value

bit number:	b7	b6	b5	b4	b3	b2	b1	b0
values	1/0	1/0	1/0	1/0	1/0	1/0	1/0	1/0
contribution	2^7	2^6	2^5	2^4	2^3	2^2	2^1	2^0
ie.	128	64	32	16	8	4	2	1

decimal range   0 - 255

The computer is able to use bytes in a number of ways:

a) bytes can be treated as numbers and the computer can perform
   arithmetic with them.
b) two different bytes can be compared to see if they have the same
   value, or whether one has a greater value than the other.
c) bytes can be stored within the computer or read in from, or
   transmitted out to, an external device, such as a printer, a
   magnetic disk unit or a laboratory instrument. It is important to
   realise that a particular 8-bit binary code may also serve a
   variety of different purposes, representing a value, a pattern, a
   character (letter, number or symbol), a binary coded decimal (BCD
   pair of digits, a hexadecimal (hex) pair of digits, and so on.
   Some examples of the quantities represented by byte values are
   given in table 5.3 and more detailed equivalence tables are
   provided in appendix 1. Finally
d) bytes are used as the coded instructions (called machine code)
   which control the operation of the computer, so that a computer
   program is actually a string of bytes treated in a particular way
   by the computer.

Table 5.2 Examples of 8-bit binary numbers

b7	b6	b5	b4	b3	b2	b1	b0		b7	b6	b5	b4	b3	b2	b1	b0
0	0	0	0	0	1	1	1		1	1	!	1	1	1	1	1

= 4 + 2 + 1   = 128+64+32+16+8+4+2+1

= 7   = 255

Table 5.3   Typical 8-bit codes

dec	BCD	hex	char	8-bit binary
65	41	41	a	01000001
66	42	42	b	01000010
90	--	5a	z	01011010
50	34	34	2	00110010
44	--	2c	,	00101100
13	--	0d	<cr>[a]	00001101
10	--	0a	<lf>[b]	00001010

Notes:
a) carriage return character
b) line feed character

Fortunately there is no necessity to plunge into the kind of
programming mentioned in d), as the microcomputers likely to be of
most value in the laboratory are supplied with commercial software
which allows the user to program the computer using a high level
language such as BASIC. Furthermore high level languages enable the
user to treat numbers within the computer as decimal quantities,
unrestricted by the small range of values which can actually be stored
in a single byte. However, it will be necessary for us to understand
something of the structure of a computer and the manipulation of bytes
if we are to make use of the computer's facility to transfer bytes
between itself and a laboratory instrument, and we shall return to
this subject in section 5.5.

The elements of a typical microcomputer are shown schematically
in fig 5.1. The system consists of a microprocessor unit (MPU) which
controls the flow of bytes around the system and actually performs any
calculations, comparisons and byte transfers required by its program,
and some memory ICs (more LSI devices) in which the bytes are stored
for use by the MPU. The MPU is generally a 40-pin IC, of which the
6502 (produced by MOS Technology), the 6800 (Motorola), the 8080
(Intel) and the Z80 (Zilog) are probably the most numerous examples.
The actual MPU used in a micro is not of great importance unless a
good deal of byte level or low-level assembler programming is
envisaged. Its true that some MPUs can operate faster than others, but
a microcomputer's behaviour under the control of a high level language
is much more likely to be dominated by the quality of the user's
program and the efficiency of its BASIC than by the clocking frequency
of its MPU.

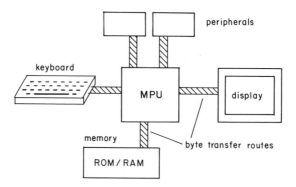

Fig 5.1   Schematic representation of the elements of a typical
          microcomputer.

Microcomputers contain at least two types of memories. First
there's read-only memory (ROM), which contains a permanently stored
program provided to control the way in which the computer operates
when switched on - and in some cases the program which accepts the
user's instructions in BASIC. Then there's read/write or random access
memory (RAM), which is used to store the user's program and data
numbers, generally for only as long as the power is switched on
(although micros with non-volatile memories, which retain their
RAM contents even after the power is switched off, are becoming more
common). A typical microcomputer may have, say, 20000 bytes (20k
bytes) of ROM containing BASIC among other things, and 16k bytes of
RAM (although a "k" (kilobyte) is actually 1024 (2^{10}) in this
context). A program cannot alter the values of bytes stored in ROM,
but only read the values that are already there, while variables and
program stored in RAM can be changed at will.

In early microcomputers the type of RAM used was known as static
RAM. This retained its byte contents until the power was turned off.
Later systems have tended to use dynamic RAM, which retains byte
values for only a few milliseconds and so needs to be refreshed every
couple of milliseconds by the application of an electrical signal to
each bit. (The popularity of the Z80 MPU for hobby micros is due in
part to the fact that it provides these refresh signals, thereby
reducing the amount of additional circuitry required.) Dynamic RAM is
more attractive as it requires less power per byte, and allows more
bytes to be stored on a given chip area. However, from the micro
user's point of view, there is very little difference between RAM
types. Several other types of memories are available for bulk storage,
and for laboratory use a magnetic disk unit (see below) is probably
the most useful for the longer term storage of programs and data. Over
the next few years we may expect to see other "add-on" memories, such
as large non-volatile RAM and bubble memories, become less expensive
and more readily available.

All microcomputers have a keyboard which is the primary pathway
of communication (input) to the computer from the user, allowing the
user to type in a program (in BASIC for example) which is then stored
in the RAM until the computer is instructed to obey the program –
again with a command from the keyboard. Most also have some kind of
alphanumeric display device (like a calculator display) or a video
output for connection to a television or a video monitor. Some micros,
like the PET, the CBM500 and the Sirius, have a built in video
monitor. This is the primary pathway of communication (output) from
the computer to the user, for displaying results and error messages
and allowing the user to check what he types in on the keyboard.

5.2 The programming language

The majority of currently available microcomputers may be
programmed in the high level language BASIC, which generally can be
learned in a few hours by almost anybody. Some micros utilise
alternative languages such as PASCAL or FORTH, and some can be used
with structured BASICs, such as COMAL, or with FORTRAN. Each language
has both advantages and disadvantages, and supporters and critics.
(This author's preference is for FORTH.) However, it is not our
objective to survey programming languages, nor to provide a lesson in
any of them. Because of the overwhelming preponderance of BASIC as the
prime language of most available low cost micros, our discussion of
the application of microcomputers to laboratory instrumentation will
be carried out in terms of software written in BASIC, and it will be
assumed that the reader is familiar with this language.

Unfortunately it is necessary to emphasise immediately that are
many different versions of BASIC, some which use a different syntax
for a similar instruction, and others which for a given syntax produce
a different result or effect. One of the most widely used forms of
BASIC is that written by Microsoft, although even this has become
bifurcated to the extent that a program written in Microsoft BASIC on
one micro cannot be guaranteed to work on another micro equipped with
Microsoft BASIC. Thus the CBM, Apple and TRS80 micros all use BASICs
written by Microsoft, although there are differences between the
versions – the TRS80's language being a version of the powerful MBASIC
usually found on Z80 machines running CP/M. Two other widely dispersed
dialects of the BASIC language are to be found on the Sinclair ZX
range of micros (although the differences between ZX81 and ZX Spectrum
BASIC are minor, such as the different syntax of the DIM statement).
Hewlett Packard produce many different computers, but their BASICs are
different again, as are those of the Acorn and the Acorn-produced BBC
microcomputer.

In some cases the differences between dialects of BASIC are small
(eg. GOTO and GO TO may not both work, PRINT USING and PRINT AT commands
may not be available, multiple assignments – eg. A=B=C=0 – may not be
permitted, functions may not require brackets and the operator used

for exponentiation varies considerably), but in others they may be
more fundamental, requiring a considerable amount of reprogramming
time if a translation form one to the other is required. For example,
A$(3) may mean the 3rd character of string A$, or, more commonly,
element 3 of a string array A$(I), and, while most BASICs use the
functions LEFT$, MID$, etc. for string handling, the ZX series use an
X$(1 TO 5) syntax. Some BASICs have MAT functions available for
handling matricies, and some allow REPEAT....UNTIL and/or
IF...THEN....ELSE structures. Rather more confusing, while most micros
use either GET or INKEY$ to test for a key pressing on the keyboard,
and return a value whether a key has been pressed or not, the GET and
GET$ functions on the BBC microcomputer and the GET instruction on the
Apple all halt the program and wait for a key to be pressed before
continuing. Most micros handle logical arithmetic in the normal (ie.
large computer) manner, but the Apple and TRS80 both use their own
special forms of logic and do not permit simple bit manipulation in
BASIC (see section 5.7).

Most of the popular micros can be programmed at the byte level
using the machine code of the MPU. (There are differences between the
machine codes for different MPUs, but once you are familiar with one
it's not difficult to get used to any other.) While such programming
can be carried out by POKing the numerical codes directly into RAM,
this is a tedious business for all but the smallest blocks of code.
It's generally much easier to write programs using nmemonic codes
known as assembler language, as this also permits the use of names
(rather than RAM byte addresses) for variables and parts of the
program. Once an assembler language program has been written it can be
translated into the equivalent byte values using a program known as an
assembler. It generally takes longer to write, test and modify an
assembler language program than an equivalent program written in
BASIC. However, assembler language programs should run hundreds of
times faster than BASIC ones - unless, of course, the speed is limited
by an external factor, such as data transfers to a peripheral or a
laboratory instrument.

For laboratory applications it is often possible to have the best
of both worlds, writing assembler routines for small, but often used,
parts of a program (such as the routines for collecting bytes from a
laboratory instrument), while retaining the ease and flexibility of
BASIC for those parts of the systems which perform calculations and
print or display results. Mastering the elements of an assembler
language is actually not as difficult as it may at first appear. The
situation is eased by the fact that, in many cases, only a handful of
the assembler language instructions need to be used. Some micros (eg.
the Acorn Electron and BBC micros) allow assembler language
instructions to be written directly into BASIC programs, while others
require that assembler language routines are loaded separately or the
appropriate codes POKEd into RAM. Most allow the use of machine code

subroutines, where a BASIC program transfers control to a machine code routine using a CALL or SYS instruction (eg. SYS 32070), with its parameter pointing to the address of the machine code routine. The machine code routine instructions are then carried out and control returned to the BASIC program by an RTS instruction, the assembler language equivalent of RETURN.

Between the worlds of BASIC and assembler language lies another pathway which may lead to speed without pain. A compiler is a program which translates a high-level language program into a machine code program. In the days of punched cards, paper tape, FORTRAN and Algol, all programs run on "real" computers where compiled, and it was the compiled, machine code form of the program which actually carried out the calculations and produced the results. BASIC changed all that. A BASIC program sits in the modern microcomputer's RAM and is interpretted BASIC instruction by BASIC instruction as it runs. Thus in a FOR loop, the BASIC instructions in the loop are interpretted each time the program cycles through the loop. Compiling a BASIC program converts it once and for all into a machine code form, so there is no interpretation while the program is running. In principle then the compiled program should run much faster than its BASIC form, although not as fast as a program written in assembler in the first place.

This author's experience of compilers for some low-cost microcomputers has been disappointing, particularly when their performance is judged against the advertising claims. In some cases the speed increase over normal BASIC has been small (eg. 2. One magazine review of a BASIC compiler noted that some programs tested actually ran slower when compiled!). For commercial use one of the attractions of a compiler is that it produces relatively tamper-proof code, and speed may be a less important aspect of the compiler's operation. In many cases a good deal of modification to (working) BASIC programs may be required before they will compile successfully. However, compilers are available for a number of the best selling micros (eg. PET, Apple, Tandy etc.) and for CP/M (see below) users, who have access to the excellent Microsoft range of BASIC compilers such as CBASIC and MBASIC (the latter being compatible with the MBASIC interpreter). An Integer BASIC compiler for the PET, although as its name implies only able to deal with integer numbers, is especially useful for compiling routines for the control of byte transfer interfaces. The compiled routines run 100 to 200 times faster than BASIC and yet the transfer of data back to BASIC for floating point calculations is quite simple. Another particulary good system available for PET and Commodore 64 users is PETSPEED, which does not actually compile into machine code, but produces a special "speed-code" which runs 20 to 40 times faster than BASIC. (The Integer BASIC compiler and PETSPEED are both produced by Oxford Computer Systems (Software) Ltd.)

5.3 The operating system

While microcomputers are programmed in popular languages such as
BASIC, they are all capable of carrying out operations which are not
part of the high level programming language. Loading and saving
programs on magnetic tape or disk, editing programs or files and
printing characters on the screen are examples of such functions.
These aspects of a micro's operation are handled by a special program
called the operating system. In some cases the operating system is
almost invisble to the user (and referred to as "transparent") - in
fact it is possible to use a micro without being aware of any
distinction between the operating system and the principal programming
language. However, moving from one micro to another can make one
uncomfortably aware of such a distinction, particulary when the new
operating system is both non-transparent and poorly designed.

The appearance of an operating system to the user (ie. what you
actually have to type on the computer's key board to make the machine
do something) varies widely between micros. Many have a system which
is unique to that model of computer, while others have adopted a
standardised operating system called CP/M (which may stand for Control
Program for Microcomputers, although opinions about this differ). CP/M
is normally loaded from disk and used in RAM, as are the various
languages available for CP/M machines. Micros using this operating
system are generally Z80 based and have (almost) a full 64k of RAM
with only a small amount of ROM. This does not actually give the user
much more memory than a 32k PET or BBC micro, or a 48k Apple II, as
much of the RAM is taken up by the CP/M and language software.

The computers most likely to be encountered in a laboratory
context tend to have unique operating systems, so there is little at
present to be gained from considering any one particular system.
However, it is well worth the effort to examine several different
machines when considering the purchase of a micro for laboratory use,
paying special attention to the use of the operating system. A good
operating system can make an enormous difference to the ease with
which a laboratory system can be programmed, tested and operated. Most
micro users tend to feel that the system they are most familiar with
is the "best" system, but in practice all have some merits and some
drawbacks. For example, the CBM range of micros has probably the best
editor for correcting and modifying programs in memory, but some have
no facility for recovering from a machine code loop (a type of
programming blunder which leaves the machine incapable of doing
anything) other than turning off the power and losing the memory
contents. The Acorn Electron and BBC computers allow assembler code to
be incorporated in BASIC programs (a useful facility where interfacing
with laboratory instrumentation is involved), but have rather old
fashioned editing facilities.

In choosing a microcomputer for relatively fast applications (eg.

where thousands of byte transfers per second are envisaged), or for applications which are likely to require the manipulation of thousands of bytes, it is wise to ensure that the micro's operating system is well documented. This allows the user to find the information required to use the system's built-in machine code routines for some aspects of data handling, rather than writing his own in machine code or relying on relatively slow BASIC for everything. For some micros this information is widely available (eg. for the PET in "The PET Revealed", "Programming the PET/CBM" or "PET machine language guide"). For others the documentation is often exchanged through users' clubs, and any serious micro user is bound to benefit from joining one of these. Generally the value of documentation provided by the manufacturer ranges from fair to poor.

5.4 Peripherals

Nearly all of the major microcomputer manufacturers supply a range of peripheral devices which may be attached to their micros. Printers are particularly valuable for the production of a "hard copy" of results, although they are also extremely useful when one is working on program development – along with a massive stock of paper to allow for many listings of the program before all the bugs are ironed out. Printers are also available from a number of independent manufacturers, and often these are cheaper or of better quality than otherwise similar printers supplied by the micro's manufacturer.

Daisy wheel printers function like typewriters, physically imprinting specific characters onto paper one after the other. Dot matrix printers produce their characters in the form of patterns of dots created by a column of (usually) 5, 7 or 9 "wires" striking the ribbon, or developing heat sensitive or electrosensitive paper. While the print quality of dot matrix printers is not generally as high as that from the daisy wheel printers (and there are exceptions – the Saunders Technology S700 Varioprinter is a dot matrix printer capable of excellent quality), dot matrix printers are probably more useful for laboratory applications. They tend to be faster than daisy wheels, and many are capable of producing "high resolution" dot graphics (albeit rather slowly), allowing hard copy of graphs, histograms, spectra etc. to be recorded.

However, one should be cautious about purchasing a printer not specifically intended for use with a particular micro, as the interfacing between micro and printer is not always straightforward. The most common hardware interfacing systems are described in chapter 7, but it is the software interfacing which can give rise to major problems where printers are concerned. For example, if one wishes to LIST a program on the printer which is plugged into the back of the micro, one needs to be sure that it is possible for the micro's operating system to send the characters out of that particular connector. (Thus all output can be diverted to the I/O slots of an

Apple, but neither the PET/CBM nor the BBC machines can easily output
a listing through their user ports.) Also some micros use reversed or
graphics characters in screen listings to represent control
instructions within strings; what does the printer make of these? The
PET range of micros use a different character code from that use by
most printers, so when upper and lower case printing is required
either a code conversion routine has to provided by the user's
program, or a hardware converter must be used. If you happen to be
planning a purchase perhaps the best approach is to ensure that you
get a demonstration of the micro and printer working together and
under your instructions before buying the printer.

   Another very valuable peripheral is the floppy disk unit. A
floppy disk is a magnetic storage medium on which programs and data
can be stored for long periods (ie. years), and yet copied back into
the computer in a few seconds. Floppy disks come in several sizes
(approximately 3, 5 and 8 inch diameter), and data can be stored on
them in a variety of ways - most computer manufacturers naturally
choosing a different storage density and layout to ensure complete
incompatability with any competitor's product. The units which hold
the floppy disks and arrange the recording and playback of information
are normally supplied to hold one or two disks, and may record on one
or on both sides of the disk. The disk itself is positioned in the
drive unit when required, and removed and stored in a dustproof sleeve
when not in use.

   Most disks can hold more information than the memories of the
popular low cost micros under consideration here. As a result they
tend to be used for the storage of quantities of data which would be
too large to keep in the computer, or which are too valuable to keep
in the computer's memory alone - where they may be at the mercy of a
passing button pusher or the vaguaries of the electricity supply
company. This in turn suggests that the greater the storage capacity
of the disk the better. But there is an alternative point of view.
While disks are remarkably reliable they may eventually fail - through
scratching, loss of magnetic coating, accidental erasure or physical
accident - and the data on them may be irretreivably lost. (In the
author's laboratory a pack of ten disks was purchased with a new disk
unit and most were in frequent use for a period of two years, during
which time not a single failure occurred. Then, within a period of two
weeks, three of the disks failed.) Thus a golden rule for the disk
user is "always keep a backup copy of anything you would hate to
loose". This means copying the content of a disk onto a spare disk. If
data is being added to continually this may require daily copying of a
disk's contents - although strictly speaking it is only the latest
data which should need copying to a backup disk - and for really
valuable data a second, perhaps less frequent, backup copy may also be
advisable. The difficulty is that copying a disk takes time, and the
larger the amount of information on the disk the longer the time
required for taking backup copies. Small capacity disks (eg. 100

kbyte) may take a couple of minutes to backup with a dual disk unit,
while a 1 Megabyte disk may take 10 minutes. Its not so much that the
time is a problem, but that some people tend to forget to take long
backups more readily than they forget to take short backups.

Before leaving the subject of disk units a word about their
operating systems. Generally the user does not have to concern himself
with where on the magnetic surface particular items of data are
stored, the information being read from and written to disk "files"
with specified filenames. It is the disk unit's controlling software
(called the Disk Operating System, DOS) which decides where to place
or find information required by the computer. Some manufacturers have
the disk unit's operating system stored and used by electronics within
the disk unit, and the unit is called an "intelligent" unit. Commodore
do this, for example, and the result is that the operation of their
disk units is virtually independent of the operation of the computer.
However, other manufactures have the disk unit's operating system
within the computer's memory - all signals required for placing and
finding data on the magnetic disk being transferred from the computer
to the disk unit, and the latter being unable to operate without
control from the computer. The disk operating system thus occupies
part of the computer's memory (and may require more than a 10k of
that), so that the memory space available to the user is reduced. The
CP/M system functions in this manner. The BBC microcomputer utilises
DFS (the F is for filing) which is held in an accessory ROM, which is
convenient as it allows independent suppliers to produce their own DFS
ROMs.

Every microcomputer has some kind of facility for storing
programs and data on a much cheaper magnetic storage medium - magnetic
tape. The majority will interface to low-cost domestic cassette
recorders through the recorder's microphone and earphone sockets,
although some makes of recorder distort the signal level too much for
this technique to work properly. Other micros require either specially
modified cassette recorders (in the case of the PET, for example, the
modification described in "The PET revealed" consists of a couple of
components costing pennies), or non-domestic cassette systems. Only a
few of the latter allow rapid searching through a tape for a specified
filename, so that in general tape must be regarded as a relatively
slow mass storage medium compared with floppy disks. A number of
companies offer enhanced tape systems, using fast moving cassettes or
video cassettes, but the costs tend to remind one of the
attractiveness of floppy disks. For the Spectrum the tape loop
Microdrives offer a remarkably low cost storage medium with many of
the characteristics of small capacity floppy disks.

Hard disks (sealed units in which the disk spins much more
rapidly than in the floppy system), also known as Winchesters, are
becoming available at reasonable prices (ie a few thousand dollars).
These can hold much larger quantities of data than floppies (eg. 3 -

30 Megabytes), so keeping backups becomes more of an art. They are useful if several micros can be connected to the same hard disk, or if a particular laboratory application requires the storage and recall of really massive amounts of data (eg. pattern matching of spectra or chromatograms).

In the last couple of years there has been a dramatic growth in the number of graphics plotters available for connection to micros. One or two rather well known companies have produced such instruments for many years, but now there is a range of reasonably priced alternatives(eg. $500 - 1000 from companies such as Watanabe and JJ Lloyd Instruments). These devices enable one to program the computer to draw virtually anything, and to add printed characters. Essentially the program gives the plotter two sets of x and y coordinates and the plotter draws a line between them. The resolution of available plotters (ie. the smallest movement the pen can make) varies from about 0.01 to 0.5 mm, and plotters are available for paper sizes from A4 (approximately 8*12 inches) up to A2 (approximately 16*24 inches). Some plotters are quite intelligent and can be instructed to draw smooth curves through a number of data points, draw and annotate axes of graphs, or even change pen colours during operation. Others rely on computer software for plotting under the control of a BASIC program. In this case the necessary routines are generally stored in a plug-in ROM supplied by the plotter's manufacture, and may not be useable with compiled BASIC.

Some of the more expensive plotters will allow the user to position the pen manually and then transfer the x and y coordinates of the pen into the computer. This technique is useful for "digitizing" data from a graphical form (eg. old spectra, chromatograms or diagrams). For some micros (eg. the Apple, PET and BBC models) digitizing "tablets" are available to offer this fuction without the plotting facility, the user tracing over the required diagram with a special pen while the diagram is held flat on the tablet's surface.

5.5 Byte handling busses

Of fundamental importance in understanding the operation of the computer is the structure of the byte handling pathways or "busses" within the machine. Figure 5.2 shows the way in which the MPU, ROM, RAM and byte input/output system are interconnected. The MPU is actually connected to two 8-bit (ie. 8 conductor) busses, one is the data bus along which bytes are transferred to or from the MPU, and the other is the control bus which the MPU uses to control other parts of the microcomputer. The duration of the digital signals is controlled by a clock signal derived from an oscillator and running at a frequency typically in the range 1 - 8 MHz. One of the signals generally available on the control bus is a clock signal of some kind, although usually a modified form of the MPU's clock input signal. There is also a 16-bit bus, called the address bus, on which 16-bit

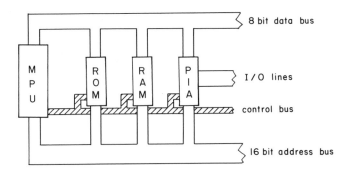

Fig 5.2   The principal byte handling pathways and components of a
          typical microcomputer.

binary numbers are planted by the MPU. An address is used primarily to
specify which byte of ROM or RAM is being referred to by the MPU.
Where a 16-bit address bus is used addresses can range from 0 to $2^{16}-1$
(65535 otherwise known as 64k), so that the MPU can address up to
65535 bytes. Typically the RAM may be connected to addresses 0 - 32k,
and the ROM to addresses 35k - 55k, other addresses being used for
other purposes. When programming in BASIC the user does not have to
worry about which bytes are being used to store data. However, a
particular byte of memory can be referred to in a BASIC program, as
illustrated in fig 5.3, usually (in Microsoft BASIC) by using the
instructions Y=PEEK(X), which collects the number (data) stored in the
byte addressed as X and stores it into Y, and POKE X,Y, which stores
the data value held in Y into the byte with the address X. (Of course,
POKing ROM doesn't work.) The BBC microcomputer's version of BASIC
uses the query indirection operator (?) for the equivalent of both
PEEKing and POKing, viz. Y=?X and ?X=Y.

Most microcomputers have additional systems connected to the
three busses, and the most important in the present context is the
peripheral interface adapter (PIA) circuit, (appearing sometimes as
peripheral input/cutput (PIO) or as the more sophisticated versatile
interface adapter (VIA) circuits.). PIA circuits are used to allow the
MPU to communicate with other byte handling systems which are not
synchronised with the computer's clock and which could give rise to
timing difficulties if the data bus itself was used as the
communication channel. The keyboard, printers and magnetic tape
recorders are examples of systems often connected to PIA circuits.
While a real PIA system is quite complex, for our purposes it is
sufficient to regard it as a circuit which allows electrical
connections with the computer's byte handling data bus.

A most useful connection provided on a number of microcomputers
is a "memory-mapped" parallel port, often controlled by a PIA circuit

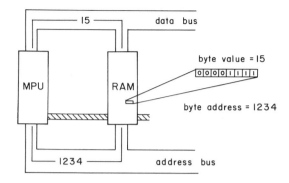

Fig 5.3   Referring to a specific byte within the micro's memory from
          BASIC. In most BASICs the function PEEK(1234) returns the
          value of the byte, while the instruction POKE 1234,15 stores
          the value 15 into the byte.

and consisting of an multi-pin connector which can carry the eight
parallel digital signal levels which make up an 8-bit byte of data,
and one or two other signals used for control purposes. A typical
example is illustrated simply in fig 5.4, although we shall examine a
real port system in more detail later in this chapter. The advantage
of such a system is that we can input the byte value of eight hi or lo
logic level signals present on the port wires, or output eight hi or
lo logic levels, using a BASIC program, by treating the 8 wires as
though they were a byte of memory (hence the term "memory-mapped") and
addressing that byte from the program. Some computers use specific
instructions (such as IN and OUT) to transfer bytes in and out of
parallel ports, while with others we may utilise the PEEK and POKE
type of instructions we met in addressing memory bytes. We shall
return to the use of this byte transfer technique shortly. As single
IC PIA and VIA circuits compatible with most of the popular MPUs are
manufactured, micros which are not fitted with a (useable) PIA can
generally have one fitted with little difficulty. In some cases "plug-
in" boards are available for connection to some convenient point on
the micro's circuit board (eg. the I/O slots in the Apple).

     A second connection system provided on many micros is a standard
input/output interface connection, some types of which we shall be
discussing in chapter 7. Popular connection systems include the IEEE
488 standard system, the Centronics[R] parallel standard port (which
provides output only) and the RS232C-type standard system (appearing
as RS232C or, in a modified form, as RS423). With the exception of the
IEEE 488 system, these connections are provided primarily to enable
the micro to be connected to computer peripherals, such as printers
and plotters, rather than laboratory instrumentation. Nevertheless the
RS232C standard is becoming increasingly popular as standard

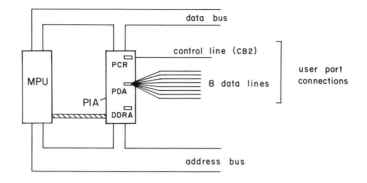

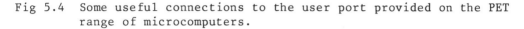

Fig 5.4   Some useful connections to the user port provided on the PET
           range of microcomputers.

connection system for laboratory instruments, and so is providing an
increasingly used route for the transfer of data between such
instruments and a microcomputer. The principal advantage of these
kinds of connections is that their use from within a BASIC program may
be much simpler than the alternative, less standard interfacing
systems we meet in chapter 6. The price paid for such convenience is
generally a cash one, but if a system is complex and expensive the
cost of a standard I/O interface is probably worth paying.

5.6 The video display

     The last of the major connections to the micro which we shall
consider is that provided to enable the computer to produce an output
display. Virtually all microcomputers currently available are designed
to produce their display as a raster scan on a video monitor or TV, as
illustrated in fig 5.5a. (There is an alternative approach, known as
vector imaging, but this is at present confined to the more expensive
end of the small computer market.) The exact method adopted for
generating the display varies rather widely and is unlikely to have a
direct impact on the laboratory use of a micro. Consequently we shall
discuss the topic in a general way and ignore the details of the
relationship of the video output system to the rest of the byte
handling circuitry.

     Standard video signals consist of three signals mixed together: a
vertical synchronisation (sync) pulse, the end of which signals the
start of scan from the top of screen; a horizontal sync pulse, which
signals the start of scan from the left of each scan line; and the
intensity level signal, which determines the beam intensity - and so
the brightness of the spot on the screen. The intensity signal is a
voltage level between 0.5V (for black) and 1.5-2V (for white - or
maximum brightness of the screen's phosphor colour), while the sync

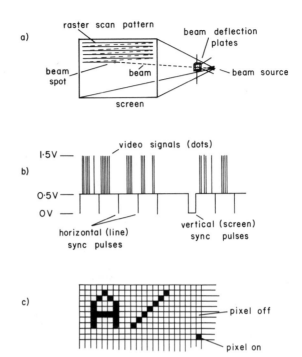

Fig 5.5   (a) The elements of the raster scan video display system. (b)
          Typical video signals. (c) The formation of a typical
          character from pixels.

pulses are drops from the 0.5V level to 0V. When all three signals are
combined together the signal is called composite video, an example of
which is shown in fig 5.5b, and this is suitable for direct connection
to a video monitor. When the video signal is used to modulate an ac
signal of the appropriate (UHF) frequency the resultant signal may be
used to drive a normal television receiver.

   A raster scan produces images on the screen by varying the
intensity of the scanning electron beam, and in microcomputer
applications the intensity variation allowed is only on or off. The
on/off condition is determined by the values of a succession of logic
pulses, representing 0s or 1s but scaled to the video signal levels,
each hi level pulse producing a single dot on the screen. For the
output of a character the video signals are formed in such a way that
the pattern of dots will appear on the video screen at a precisely
determined time after the start of the electron beam scan, and
therefore at a precisely determined posistion on the screen.

   While some micros provide composite video signals, most use the
video signal to modulate a UHF waveform so that the modulated signal

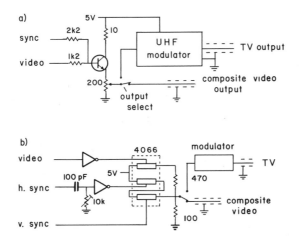

Fig 5.6   a) Extraction of composite video signal from the modulator
          input of typical home computer. b) The generation of
          composite video from video and sync signals available on PET
          user port.

is suitable for use with a standard domestic television. Televisions
do not have the stability or the bandwidth (ie. white/black transition
speed) of good quality video monitors, so a computer display viewed on
a TV screen never looks as good as one on a proper monitor - the dots
appear fuzzy on a TV and the total picture tends to wander around. For
most laboratory applications one can live with these problems,
although modifying a micro to bypass its UHF modulator is not
difficult and can be done by breaking the modulators input lead and
taking the composite video signal from the mixer, as illustrated in
fig 5.6a. (Of course, this probably voids the computer's warranty!) In
the case of the PET range of computers the video intensity and sync
signals are available separately on the computer's user port and may
be mixed to produce composite video (using a circuit such as that
shown in fig 5.6b) for display on an external monitor.

    Many popular micros now produce signals for multicolour display
on colour televisions, and these originate from three different
intensity signals, red, green and blue. High quality monitors for
colour displays are called RGB monitors and have separate inputs for
the three colours intensities and one for the sync signals. Some
micros (eg. the BBCs) provide RGB connectors. Mixing the colour levels
in the proportions of 1:2:4 (for blue, green and red respectively) to
generate colour video levels, and adding in the sync pulses, produces
the colour equivalent of the composite video signal. This may be used
to drive the single input colour monitors associated with CCTV and
audio-visual systems, and is a facility provided on an increasing
number of micros. In most cases this signal is modulated in the same

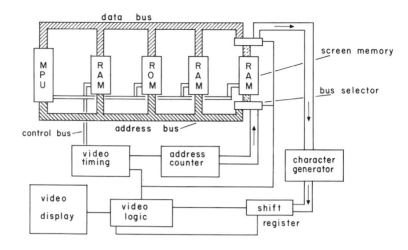

Fig 5.7 The video display of many micros uses a section of RAM known
         as screen memory, the contents of which are used to generate
         characters in pixel matrix form.

way as for the monochrome systems, and the composite video signal may
be extracted from the emitter load of the mixer transistor in the
manner shown in fig 5.6. (Note: If the emitter resistance is not
variable it is advisable to include a small value, eg. 75 R, fixed
resistor in the video output signal line to protect against
overloading the monitor input.)

     There are three broad approaches to the generation of video
signals for a screen display. The first is character position mapping
illustrated for a monochrome system by the block diagram illustration
in fig 5.7. The central feature is a block of RAM (screen memory) into
which the computer's operating system places bytes representing the
characters which will be displayed on the video screen in response to
PRINT instructions etc. This block of RAM is also being frequently
examined (typically 50/60 times a second) by a second hardware system
generally referred to as the video logic system. This system extracts
from the screen RAM each byte in turn (at times when the RAM is not
being used by the MPU), and passes it to a pulse generating circuit
often made up of a character generator ROM connected to a shift
register. This circuit translates the byte code into a serial pattern
of pulses which are then converted into video signals, forming a
pattern of dots on the video screen.

     The screen display is generally arranged as a rectangular matrix
varying from 16 lines of 20 characters positions (for TV display
systems) to 25 lines of 80 characters positions (for the more
expensive video monitor systems). As at least one byte of RAM is

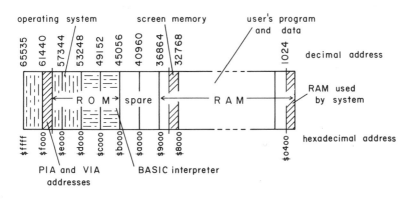

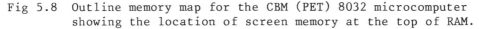

Fig 5.8   Outline memory map for the CBM (PET) 8032 microcomputer
          showing the location of screen memory at the top of RAM.

required for each character position, the "screen memory" may occupy
anything from a few hundred to 2k bytes of RAM. The addresses used for
screen memory may be found in the microcomputer's memory map. A
typical example (for the CBM 8032 computer) is given in fig 5.8, where
the screen memory occupies addresses 32768 to 34767 inclusive. If the
micro's operating systems allow direct access to the screen memory,
then the contents of character positions may be examined or changed
using PEEK and POKE instructions repectively. Otherwise the screen
display is produced by simple PRINT statements within the user's
program and a variety of operating systems functions (such as LIST
etc.).

     Each screen character position has its dot pattern determined by
the contents of one byte (in single colour systems) of screen memory.
Each byte can hold values in the range 0 - 255, although less than
half of these are required as codes for all the upper and lower case
letters, numbers and punctuation marks. As a result some micros use
the "spare" values to provide dot patterns which are useful for
generating graphic displays on the screen. The PET, VIC, Spectrum and
Apple computers, for example, provide for a range of graphics
characters, such as lines, squares, bars and in some cases the card
suite symbols. These graphics characters allow the user to build up
simple but crude pictures on the screen to represent graphs and
histograms, etc. The detail which can be included in such pictures is
small because of the limitation in the number of character positions,
although for most laboratory applications pictures on screens are not
particulary valuable - hard copy obtained on a graph plotter is more
likely to be of serious use.

     A second character based display system has developed from the
ROM generator technique described above. Some micros (eg. the
Commodore 64 and VIC-20) allow bit patterns for character generation

to be stored in RAM, so that the dot patterns defined for byte value
in screen memory may be changed by programming. The approach adds
considerably to the range of shapes which may be produced on the
display screen, although retaining the concept of character position
within the display.

A third display system, known as the bit mapped display, uses a
bit in RAM to represent every individual dot position on a screen. F
example, when characters are displayed on the Spectrum they are
treated as 8*8 dot matricies, requiring 8 bytes of RAM to store the
pattern for a single character. For the top display line (of 32
characters on the Spectrum) the 32 bytes representing the top 8 dots
for each character, 256 bits, are stored consecutively, followed by
bytes representing the top 8 dots of each character of the second
line, and so on for the first eight lines of the display. Then come
the bytes holding the second row of dots for the characters in the
first 8 lines, then the third rows, etc. When the 64 rows of dots fo
the first 8 lines of characters are complete we find the dot pattern
for the second 8 lines of the display, and finally the third 8 lines
This approach requires 6144 bytes of RAM for the 49,152 dot position
plus 768 bytes for the attributes (the colours, whether flashing,
etc.) of each 8*8 character position. It also precludes the use of
PEEK and POKE for the display memory, as a byte now only holds
information about one row of the dots forming a complete character.
(SCREEN$ and PRINT AT are used to access display positions instead).
the other hand it does allow dot graphics (usually called high
resolution graphics) to be easily used from BASIC (with PLOT and POI
instruction) and makes it easier to use non-standard characters
anywhere in the display, not just at the character positions of the
earlier display systems.

There are many applications for which high resolution graphic
displays are desirable (eg. pattern matching, viewing of spectra or
chromatograms, etc.) and most micros can generate such displays by
utilising the individual dots (called pixels) of which the normal
character displays are formed. (The PET micros are not equipped to
produce high resolution graphics, but a number of companies market
add-on systems for providing this facility, and these have the
advantage of not using up the PET's RAM.) High resolution graphic
systems enable the user to set any pixel of the display either off
(black) or on (white), by setting the individual bits of bytes in RA
to 0s or 1s. Displays range from, say, 120 * 200 pixels to 400 * 600
pixels, so high resolution graphics tends to use rather a lot of RAM
although in most cases (and unlike a fixed character display memory)
the memory is not dedicated to this function, so it's available for
other purposes when the high resolution graphics capability is not i
use. Most systems are provided with the software to enable simple
operations, such as drawing lines and filling in areas, to be
accomplished relatively easily from within a BASIC program. Typicall
the requirement on the user's program is to provide x and y

Table 5.4   The AND operation and its implementation in BASIC

The bytewise effect	BASIC examples

```
The bytewise effect              BASIC examples
_____

X                    15          IF (X AND 1)>0 THEN 100
----------------------               goes to 100 if b0=1
0  0  0  0  1  1  1  1
----------------------           Z = X AND 64
Y                   170          IF Z = 0 THEN GOTO 2000
----------------------               goes to 2000 if b6=0
1  0  1  0  1  0  1  0
----------------------           L = X AND 15
X AND Y              10          H = X AND 240
----------------------               L & H are nybbles of X
0  0  0  0  1  0  1  0
----------------------           IF X >= 128 THEN 1000
(15 AND 170) results in  10          goes to 100 if b7=1
```

coordinates (in pixel numbers) for a single point, or for the start
and end points of a line for a simple BASIC-like instruction.

   High resolution images on video monitors can be photographed
fairly easily, although the quality of the image on a standard TV
screen is rarely adequate. One advantage of high resolution graphic
displays is that the composite video signal used to generate the
display may be fed to a device called a video copier, which at the
touch of a button (or command from the computer) can produce a paper
photocopy of the pixel pattern on the screen - thus allowing the rapid
production of hard copy (although, generally, of lower quality than
could be obtained from a slower pen plotter).

5.7 Bit manipulation

   High resolution graphics and several aspects of the byte handling
systems we shall discuss below make use of the rapid and simple
setting or testing of particular bits of a byte. For example, parallel
port connectors generally have at least one control line output in
addition to the eight data lines, and it is necessary to be able to
switch the level of this line between hi and lo to operate a
peripheral connected to the port. In many cases the control line is
actually connected so that it resembles one bit (or more) of a
specific byte associated with the PIA circuit, so that from the
programming point of view we need to be able to change the value of
one bit of that byte while leaving the other seven bits unchanged.

Table 5.5   The OR operation

The bytewise effect   BASIC examples

X   15
----------------------   Y = X OR 128
0 0 0 0 1 1 1 1   set b7 of Y to 1

Y   170
----------------------   IF (A OR B)=1 THEN 100
1 0 1 0 1 0 1 0   go to 100 if b0 of A
----------------------   or b0 of B = 1
X OR Y   175
----------------------   Y = X OR 32
1 0 1 0 1 1 1 1   set b4 of Y to 1,
----------------------   rest of Y as X

15 OR 170   results in   175

Testing or changing specific bits of a byte may be accomplished
on most micros in BASIC with the aid of Boolean operations such as AN
and OR . Such operations involve comparisons between corresponding
bits of two bytes and produce a result which is stored in a third
byte. For the AND operation, illustrated in table 5.4, the result of
the BASIC instruction Z = X AND Y is found by setting each bit of Z t
a 1 if the corresponding bits of both X and Y are 1s. Those bits of Z
for which the corresponding bits of X and Y are not both 1s are set
equal to 0s. The AND operation is most frequently used to pick out
particular bits of one byte by ANDing with a second byte in which the
bits of interest are 1s. Thus to discover whether bit 0 of X is a 1 w
could take X AND 1 (since decimal 1 is 00000001) and the result would
be zero if b0 of X was a 0, but non-zero if b0 of X was a 1. Some
samples of BASIC coding for testing or extracting bits of X are
included in table 5.4. The AND operation is also useful for setting
particular bits of a byte to 0 without effecting other bits of the
byte. For example X = X AND 127 sets b7 of X to a 0 without effecting
the value of the other seven bits. (127 is 01111111 in binary).

The OR operation is illustrated in table 5.5, where the result c
the BASIC instruction Z = X OR Y is found by setting each bit of Z to
a 1 if either (or both, it's an inclusive OR operation) of the
corresponding bits in X or Y are 1s. Those bits of Z for which the
corresponding bits of both X and Y are 0s are set equal to 0. The OR
operation is most commonly used to force particular bits of a byte to
be 1s without effecting the values of other bits in the byte. For
example X = X OR 128 sets b7 of X to a 1 without changing the value c

Table 5.6   BASIC subroutines for byte transfer through
            the PET user port

---

byte input subroutine: data returns in X

```
100 POKE 59459, 0 :  set port for input
110 POKE 59468,PEEK(59468) AND 31 OR 192
        switch control line to 0V
120 X=PEEK(59471) :  input data and store in X
130 POKE 59468,PEEK(59468) OR 224
        switch control line to 5V
190 RETURN
```

---

byte output subroutine: uses data in Y

```
200 POKE 59459,255:  set port for output
210 POKE 59471,Y  :  output data stored in Y
220 POKE 59468,PEEK(59468) AND 31 OR 192
        switch control line to 0V
230 POKE 59468,PEEK(59468) OR 224
        switch control line to 5V
290 RETURN
```

---

any of the other seven bits. (128 is 10000000 in binary). The BBC
micro also provides an EOR operator which allows an Exclusive OR
comparison between two bytes (ie. the bitwise result is a 0 if both
bits are 1s, cf. section 4.1). Unfortunately the logical operators
provided on the Apple and TRS80 computers do not support bitwise
operations and so cannot be used in the manner described above.

The PET range of microcomputers utilise address 59471 as the 8-
bit parallel port for input and output of bytes. However, two other
addresses associated with the PIA are also used for the control of
data transfers via the parallel port (see fig 5.4). The first is
59459, which is called the data direction register and must be set to
contain a 1 in each bit for which the corresponding bit of 59471 is to
be an output line, or a zero in each bit for which the corresponding
bit of 59471 is an input. Thus if we wish to use all 8 lines of the
parallel port as inputs we must use POKE 59459,0 before we read the
port with X=PEEK(59471). The second address of interest is 59468,
called the peripheral control register, in which bits 5 to 7 are used
to determine the voltage on the control line output of the user port.
(The use of this control line will become apparent in chapter 6. For
the moment we shall assume that devices connected to the user port can
be instructed to place their data on, or remove data from, the 8 data
lines by a hi-to-lo transition on the control line.) When these three

bits are 111 the control line output is 5 V, when the bits are 110 t
output is 0 V, (any other values cause rather different effects). Th
BBC microcomputer uses the same VIA circuit (a 6522) for its user po
system, although the addresses used for communication with the VIA a
the uses made of some of the control lines are different.

We can now look at some BASIC subroutines to transfer bytes of
data to or from the parallel port of a PET computer. Table 5.6 shows
such subroutines for the input (lines 100-190) and output (lines 200
299) of a single data byte through the port. The input subroutine se
the data direction register to 0 (to configure the port for input),
changes the voltage level of the port's control line to 0 V, then
PEEKs at the port lines and stores the data byte into the BASIC
variable X before changing the control line voltage back to 5 V and
RETURNing to the main program. The output subroutine sets the data
direction register to 255 (for output on all 8 lines), POKEs the dat
byte onto the port lines, then toggles the control line, first to 0
then back to 5 V, providing an output pulse which may be used to
control the operation of a peripheral attached to the port. (Note th
31 = 0001 1111, 192 = 1100 0000, and 224 = 1110 0000.) To make use o
the CB2 line in this manner it is necessary to configure the VIA in
such a way that its internal shift register (which is connected to C
and may be used for serial I/O) is disabled. This may be achieved by
setting bits 2 - 4 of the auxilary control register (59467 on the PE
to 0s at the begining of any program using CB2 for control purposes,
eg.

10 POKE 59467, PEEK(59467) AND 227

Further details of the VIA and its operation are provided in The PET
Revealed and in Programming the PET/CBM.

5.8 Timing and addressing

The details of the operation of the MPU are beyond the scope of
this text; several sources of such information are included in the
bibliography. However, some aspects of interfacing (dealt with in
chapters 6 & 7) require a somewhat deeper appreciation of the contro
of byte transfers over the data bus than has been necessary so far.
Here we introduce in outline the manner in which the MPU reads bytes
from, and writes bytes to, a specified address in RAM, as a prelude
our later discussion of reading and writing bytes at specified
addresses associated with laboratory interfaces.

Every operation within the MPU is carried out on a specified
transition of its clock input, ie. when the level on the clock input
is changing from hi to lo or lo to hi. Consequently, collecting
addresses from the address bus and placing data bytes on, or
collecting data bytes from, the data bus requires that other circuit
on the bus (eg. ROM, RAM, PIA) are synchronised with the MPU. This

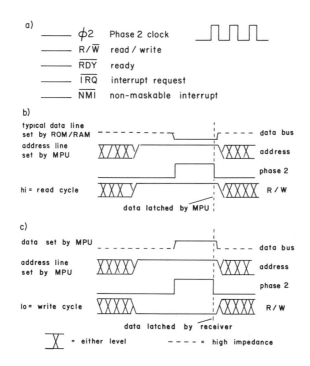

Fig 5.9 a) Some of the control bus lines used by MPUs which control
          their read/write operations through a R/W line and a clocking
          signal. The synchronisation of the b) read, and c) write
          operations. The cross hatching indicates that the levels on
          the lines are irrelevant.

synchronisation is achieved by signals on the control bus. The precise
details of the control bus signals vary with the MPU, but most fall
into one of the two main categories, illustrated in part in figs 5.9 &
5.10.

    In fig 5.9a we see the principal control bus lines and signals of
MPUs such as the 6502 and 6800. Operation may be understood by
examining the "read byte from memory" and "write byte to memory"
sequences illustrated in fig 5.9b and c. The data bus, being a TRI-
STATE bus, is normally in the high impedance state. When a byte is to
be read from memory the MPU places the byte's address on the address
bus and holds the read/write (R/W) control bus line hi to signify a
read operation (ie. the addressed device has to place data on the data
bus). Now nothing actually happens anywhere else until a second
control bus line called phase 2 goes hi. Phase 2 is actually a
clocking signal closely related to the MPUs clock input, although
sometimes it's a non-symmetrical signal. When phase 2 goes hi the

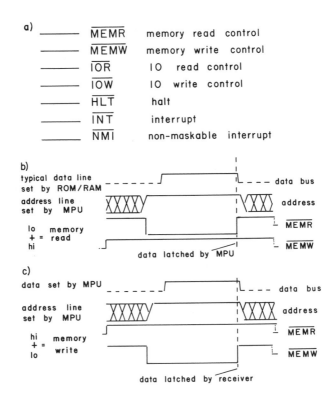

a)
	MEMR	memory read control
	MEMW	memory write control
	IOR	IO read control
	IOW	IO write control
	HLT	halt
	INT	interrupt
	NMI	non-maskable interrupt

Fig 5.10 a) Some of the control bus lines used by MPUs which control
their read/write operations through separate MEMR and MEMW
lines. The synchronisation of the b) read, and c) write
operations. The cross hatching indicates that the levels on
the lines are irrelevant.

addressed device (say a ROM) is supposed to respond by placing the
value of the addressed byte onto the data bus. As this process can
take one or two hundred nanoseconds before the data bus lines
stabilise to the required his and los, the MPU waits until phase 2
returns to lo. This hi to lo transition corresponds to the instant
when the MPU latches the data from the data bus into one of its own
buffers or registers (ie. it has read the data byte), and at this
point the address and data bus lines and R/W are freed for use in the
next instruction (the data lines, being TRI-STATE, return to the high
impedance state).

A similar sequence of events governs the writing of a byte to
memory, as illustrated in fig 5.9c. The MPU sets the address line to
the required values and sets R/W lo to indicate a write operation. The
rise of phase 2 from lo to hi coincides with the MPU placing a data
byte on the data bus. That data byte is stable by the time that the

phase 2 line returns to lo, and this hi to lo transition is often used
to latch the data line values into RAM or the PIA. Shortly after the
transition the MPU releases the address lines and R/W, and the data
lines become high impedance. (Note: the situation is slightly more
complex in the case of the 6800 because a "valid memory address", VMA,
control line is also used.)

The key feature of the technique illustrated in fig 5.9 is that a
R/W control line is used to indicate whether a read or write operation
is required, and the phase 2 clock pulse edges are used to provide the
timing signals. The other major read/write technique uses one control
line to control read operations and another control line to control
write operations. This approach is illustrated in fig 5.10, and is
used by the 8080 MPU and (in a modified form) by the Z80. Again the
MPU places the address information on the address bus, but now either
the normally hi MEMR (memory read) is taken lo to indicate a read
operation (fig 5.10b), or a normally hi MEMW (memory write) line is
taken lo to indicate a write operation (fig 5.10c). In each case the
latching of the data, either by the MPU or by RAM, occurs when the
MEMR or MEMW undergoes a lo to hi transition.

Of particular interest with some MPUs using the second technique
(fig 5.10) is the availability of a second pair of control lines IOR
(IO read) and IOW (IO write) which can be used to control byte
transfers between the MPU and an alternative set of addressed
locations when specific instructions are carried out. Thus while
normal load and store instructions cause addresses to be loaded and
the MEMW or MEMR lines to be toggled, instructions such as IN and OUT
cause addresses to be loaded but the IOW or IOR lines to be toggled
while the MEMx lines remain hi. The Z80 addressing method is similar
but uses one pair of control lines for RD and WR, and a second pair to
distinguish between memory access and IO access - MEMRQ and IORQ. A
consequence of these arrangements is that a micro with a full
complement of 64k bytes of memory may be programmed to carry out byte
transfers with up to 64k addresses outside memory - a potential which
has to some extent been nullified by certain hardware features in some
low-cost Z80 based micros, (such as the ZX81 and Spectrum). We shall
return to this subject in chapter 6.

The popular MPUs are provided with a 16 bit address bus and an 8
bit data bus, so that 24 pins of the MPU IC are required for byte
transfers with an addressing range of 64k, plus of course the control
signals, clock and power supply pins. Until recently it was not
possible to produce memory ICs containing such a large number of
addressable bytes; most currently used memory ICs hold 2k, 4k or 8k
bytes. Consequently the memory of most of the popular microcomputers
is made up of several separate ICs, some ROM and some RAM. For this
reason it has not been the practice to connect all 16 address lines to
every memory IC, but rather to partially decode the 16 bit address
into two parts. Typically the lower 12 bits of the bus (covering the

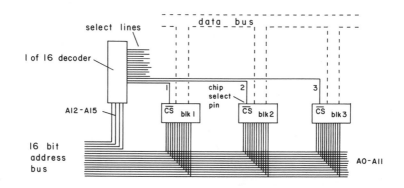

Fig 5.11 Partial decoding of a 16 bit address bus to provide a 12 bit
bus and 16 separate block select lines. In this example each
ROM block decodes the 12 bit address on A0-A11 when its chip
select line (CS) is held lo.

range 0 - 4096) are left as address lines, while the top 4 bits are
decoded to provide 16 separate signals, as illustrated in fig 5.11.
Each of these may be used to identify which one of up to 16 blocks of
4k bytes should react to the address held on the lower 12 bits of the
address bus. This technique is widely used for addressing ROM,
allowing standard 4k ROMs (in 24 pin packages) to be used in systems
containing >20k of ROM.

Most microcomputers have facilities allowing relatively
straightforward connection to their address, data and control busses,
often provided for the purpose of allowing memory expansion. In some
cases (eg. the Spectrum) a connector is avaialable on the case of the
micro. In others the connector is provided inside the case with no
obvious hole for a cable (eg. the PET), or with clearly indicated
slots to allow connection of additional circuits (eg. the I/O slots of
the Apple). In the case of the PET only address lines A0 - A11 are
provided on the "memory expansion connector", the higher order bits
having been partially decoded as described above. Ten block select
lines provide signals for the selection of the second to the eleventh
4k byte block (the other blocks being fully committed inside the PET).
On the Apple while all 16 address line are provided on the I/O slot
connectors, partially decoded address signals also appear on two of
the connectors pins, one for a given value of A4 - A15 (leaving A0 -
A3 for use), and one for a given value of A8 - A15 (leaving A0 - A7).
The Spectrum provides all the data, address and control bus signals of
its Z80 MPU at its expansion connector. The BBC microcomputer provides
a 1 MHz Extension Bus connector on which address lines A0-A7 appear,
while the data bus is gated by block select lines which also appear on
the connector and provide a communication pathway using address blocks
$fc00 (known as FRED) and $fd00 (JIM). Our only interest in connection

to the address, data and control busses is for interfacing purposes, and for that reason we defer further discussion until the latter part of chapter 6.

5.9 Interrupts and interrupt flags

The control bus of all MPUs includes at least one line which can be used to interrupt the execution of a program. (See for example, figs. 5.9a & 10a.) Many MPUs provide two different interrupt lines, although microcomputers based on these MPUs do not necessarily have both of these lines connected connected to anything. An interrupt is signalled by some device external to the MPU taking the interrupt line lo. When this happens the MPU stops execution of its program, stores the program counter register (which holds the address of the next instruction it would have carried out if it had not received the interrupt) and the status register in a part of RAM known as the stack, and then starts executing a second program. The start address of this second program is known as the interrupt vector, and is stored at a specific memory address which is normally part of ROM. For example, the interrupt request vector for the 6502 is expected to be at addresses 65534-5, $fffe-f in hexadecimal.

Microcomputers tend to rely heavily on the use of interrupts to carry out fundamental operations, such as the checking of the keyboard to discover which key has been pressed. Some are wired up so that an interrupt occur when specific external events take place (such as a key being pressed), others arrange for frequent and regular interrupts (such as 50/60 times a second). All have an interrupt servicing routine as part of their operating system, and this is pointed to by the interrupt vector. The interrupt servicing routine generally stores pointers and data associated with the program which was being executed, and then examines a number of parts of the system to see what caused the interrupt and to take whatever action is required. Once the interrupt has been serviced, the system restores the data and pointers of the original program and returns to execute that program at the point of the interruption. (A special return instruction, RTI, causes the MPU to reload its saved program counter and status registers from the stack. Note that many microcomputers use two stacks - one for the MPU, and a second for the interrupt servicing routine of the operating system.)

Those MPUs with two interrupt lines (eg. the 6502, 6800, Z80 etc.) generally allow the user to prevent or mask an interrupt on one of the lines (the INT or IRQ line). This facility allows, for example, the interrupt servicing routine to ensure that there will not be a second interrupt before it has finished servicing a first interrupt (otherwise things could get very complicated). However, the second interrupt line is not maskable (it's called the NMI line - Non-Maskable Interrupt), and an interrupt on this line is always attended to, by a second servicing routine with an address given by the NMI

vector, and which probably has to be supplied by the user. (The 6502
NMI vector is at 65530-1, $fffa-b.) Considerable care is needed in the
use of NMIs, although if they are not used by the micro's operating
system they do offer a straightforward way of connecting external
devices which need attention on demand.

MPU interrupts are not easy to handle without a good deal of
practice and assembler language programming, although their value for
detecting external events (eg. requests for attention from laboratory
instruments) is considerable. Because this book is intended primarily
for the high level language user we shall not deal further with the
use of MPU interrupts. Fortunately micros with PIA circuits usually
possess the facilities for setting a one bit flag (that's just a bit
of a byte) when a logic level transition is detected on a PIA control
line. This allows the user to program a response to external events
very much more easily than through MPU interrupts.

We have already noted that for the PET's VIA system, illustrated
in part in fig 5.12, one of the control lines (the CB2 line) may be
used as an logic level output, with its level determined by bit value
in byte 59468. A second line (CA1) may be used to respond to input
signals, although in this case only to logic level transitions, by
setting a bit (b1) of the (VIA's) flag register, byte 59469. This
particular bit is called the interrupt flag. (In fact setting the
interrupt flag to a 1 can cause an MPU interrupt if certain other
flags have also been set. However, we shall not be discussing the use
of the flag to generate MPU interrupts, but only to sense transitions
on the CA1 line.) Whether the interrupt flag is set by a lo-to-hi or
hi-to-lo transition on CA1 is determined by b0 of the peripheral
control register (59468 again). If b0 is a 1, then the interrupt flag
is set by lo-to-hi transitions on CA1, while if b0 is a 0 it's hi-to-
lo transitions that do the job.

Thus the computer can be programmed to wait in a loop until the
interrupt flag is set. A typical BASIC subroutine which serves this
purpose is:

```
1000 REM WAIT FOR INTERRUPT FLAG
1010 IF (PEEK(59468) AND 1) >0 THEN RETURN
1020 GOTO 1010
```

The subroutine returns to the calling routine once flag setting has
been detected. Note that addressing the flag register actually clears
the interrupt flag, so the flag is always cleared on exit from this
subroutine. Actually its likely that some useful processing could be
programmed in between tests of the flag.

Most PIA circuits actually have two eight bit parallel ports,
each with two control lines. The ports are independently useable,
although in the example we have discussed (of the PET's 6522 VIA) it

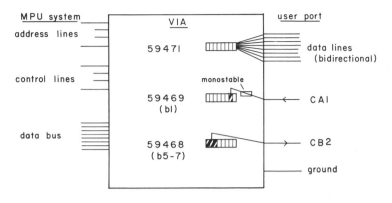

Fig 5.12 The principal connections of the VIA controlled user port of
the PET microcomputers. The VIA actually has 16 internal
registers and is used for control of some aspects of the IEEE
488 bus, the cassette interfaces and the video display.

happens that one port is used as part of the IEEE-488 interface system
and two of the control lines are utilised for cassette (CB1) and video
(CA2) purposes. For this reason its only CA1 and CB2 which are
available for external use. In general Cx2 lines may be used for logic
level outputs or as transition sensitive inputs, while Cx1 lines are
transition sensitive inputs only. On the BBC microcomputer CB1 and CB2
are available at the user port, while CA1 and CA2 are used on the
printer port.

5.10 The 16 bit micro

The enormous success of the 8 bit microprocessors encouraged the
development of 16 bit MPUs, which use 16 bit internal registers for
instruction codes and data handling and (in most cases) a 16 bit data
bus. In principle the larger capacity of the digital "words" use to
hold machine code instructions should enable a much wider range of
instructions to be employed, and a much faster execution of programs
to be achieved, than is the case with 8 bit MPUs. A considerable
number of 16 bit MPUs is now available, the most widely used varieties
at present being the 68000 (which incidentally has some 32 bit
registers, from Motorola), the 8086 (Intel) and the Z800 (Zilog), and
several of these are designed to incorporate the machine code
instruction sets of the related 8 bit MPUs - as a small subset of
their own instructions. Microcomputers based on these powerful
processors have been appearing at regular intervals, including the
Sirius, the Apple Lisa, the ACT Apricot, several Hewlett Packard and
Tetroniks models and the IBM Personal Computer. It is generally
accepted that for most 16 bit machines the software was initially
lagging behind the hardware, in that the high level language compilers
and interpreters for the processors were actually modified versions of

the software written for the 8 bit models. Consequently the speed to
be expected from a 16 bit machine was not immediately available.
However, this situation is changing and the well-known software houses
are now publishing more suitable systems.

Quite apart from the staightforward speed advantage to be offered
by (suitable) 16 bit software, 16 bit microcomputers do have two other
major advantages over their 8 bit counterparts. The first is in amount
of memory space addressable. While most 8 bit MPUs are limited by the
size of their 16 bit address bus to addressing 64 kbytes of memory, 16
bit MPUs can generally address much larger memories. In the case of
the Z800 this is achieved by the use of a 16 bit address bus plus
instructions which enable one of several 64 kbyte banks of memory to
be selected. In the case of the other processors addressing is
achieved through the (logical) use of a 20 bit (8086) or 24 bit
(68000) address bus, allowing address spaces of 1 and 16 Mbyte
respectively. Consequently the 16 bit micros have a plentiful supply
of RAM available.

The second advantage, which follows from the first, is that a
much greater quantity of RAM may be devoted to screen memory. In fact
most of the 16 bit micros with built in video monitors use a bit
mapped display in which a large number of pixels provides a high
degree of graphics resolution (eg. 800 * 400 on the Sirius).
Consequently the screen graphics capabilities of most 16 bit micros
are much more impressive than those of their 8 bit counterparts, and
holding several screen images in RAM at one time becomes a practical
proposition.

Speed, memory size and graphics capability make the 16 bit micro
very attractive for laboratory computer applications. Unfortunately
there are some disadvantages which should be considered before
purchasing one for laboratory work. In general the 16 bit micros are
more expensive than the 8 bit ones, although considering the extra
memory usually included the difference is not large - an 8 bit system
(including disks) typically costing $1000-2500, while a 16 bit system
may cost $2000-5000. The main problem (at present) is that
interfacing, and information about interfacing, is less readily
available, and the machines themselves are more complex electronically
(the address bus being multiplexed in some cases, for example). There
is no doubt that high speed data acquisition, large scale data
handling and computation (number crunching) would all benefit from the
use of 16 rather than 8 bit machines. But the fact remains that many
laboratory needs can be filled economically by 8 bit machines, and for
more exacting requirements purpose designed minicomputer systems, such
as the DEC MINC, offer a great variety of scientific software and
hardware and, a factor of considerable importance in complex systems,
the technical support required to overcome implementation
difficulties.

CHAPTER 6

INTERFACING MICROCOMPUTERS WITH LABORATORY INSTRUMENTS

We have seen that the measurement or control functions of
laboratory instrument circuits produce or utilise signals which may be
voltages, currents or pulses, while the microcomputer is a byte
oriented device which produces or uses 8-bit parallel binary signals.
To enable the signals of an instrument to form input or output data
for a computer a third device is needed to translate one type of
signal into the other. Such a device is called an interface and its
role is illustrated in fig 6.1. An instrumental signal may be a
constant value or may be time dependent, while the computer's signal
must be one or more bytes specified at a particular instant of time.
For this reason the translation carried out by the interface must be
triggered in some way, either by the computer or by the laboratory
instrument. In describing the basic types of laboratory interfaces
(section 6.1), we shall assume that the computer initiates the
translation process with a logic signal on a single control line
(using, for example, the software control described in section 5.7).
The subject will be considered in more detail in section 6.4.

6.1 Basic instrumental interface types

To accommodate the different types of instrumental signals
(input, output, analog voltages, logic levels and pulses) a number of
different types of interfaces are available. In this section we shall
briefly consider the nature of each of the common types of interfaces

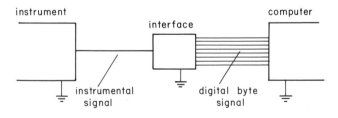

Fig 6.1   The use of an interface unit to translate laboratory signals
          into digital bytes.

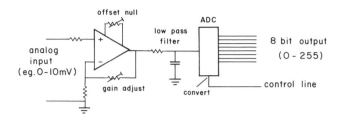

Fig 6.2   A typical analog input interface.

in turn. Most are available commercially in a form suitable for direct
connection to the more popular microcomputers, although, if high speed
and high precision are not essential, interfaces can be constructed
quite easily by a competent electronics technician.

6.1.1 Analog input interface

The analog input (or analog-to-digital conversion) interface
translates constant or slowly varying analog signals into bytes with
binary values proportional to the analog input. A typical analog
input interface is illustrated in fig 6.2, where it will be seen that
the circuit consists of two stages. The first stage is an adjustable
gain amplifier based on a non-inverting configuration op-amp to allow
a high input impedance. This converts the incoming signal from the
measurement range (say, 0 - 10 mV, the typical output range of a
chromatographic detector) to the standard input range (typically 0 -
V) of an intergrated circuit ADC. When the appropriate control signal
level is applied to the ADC chip's "convert" pin, the analog voltage
at its input (0-1V) is converted into a digital value (0-255) which is
stored in an internal buffer. The digital value can then appear as an
8-bit parallel digital signal on the output lines, although this
output is generally under the control of a signal applied to an output
disable pin or buffered through a TRI-STATE octal latch, so that the
output can be disabled when not required. By adjusting the gain and
nulling the op-amp's offset voltage, an 8-bit analog-to-digital
interface can be set to produce binary 0 output for 0 V input, binary
255 (ie, 11111111) output for 10 mV input, binary 127 for 5 mV input
and so on, (Note the quantisation error).

Integrated circuit ADCs are normally voltage driven devices, so
that if a current measurement interface is required the voltage
amplifier of fig 6.2 may be replaced by a transresistance amplifier
circuit (see section 3.5.5). Similarly if the peak value of an ac
signal is to be recorded the voltage amplifier may be replaced with a
precision diode rectifier circuit, provided that the output of this is
adequately filtered to remove ripple before the signal is applied to
the ADC input. "True RMS converters" are also available in IC form for

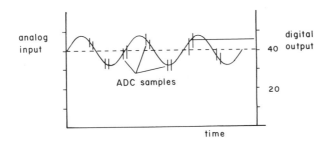

Fig 6.3   An illustration of the effects of pick-up of mains related
          interference by an analog input interface.

extracting RMS value from a variety of ac input functions, although
these are generally limited to frequencies below 1 MHz.

   The time taken by the ADC to convert the analog input into
digital output, the conversion time, varies widely with the type of
ADC (and usually its cost). ADCs are available with conversion times
in the range of less than a microsecond to several milliseconds (see
section 4.7.2). Clearly the conversion speed required in a particular
application will depend on the rapidity with which the analog signal
is changing and on how often the signal must be sampled. Analog input
interfaces are widely used for passing signals from the pen recorder
output connections of instruments (such as chromatographs and
spectrometers) to a computer, and in such cases a relatively slow ADC
is usually adequate (eg. conversion time c. 1ms). Of course,
unscreened leads which may have been adequate for connection to a
chart recorder will need to be changed to properly screened cables and
connectors for connection to an interface.

   All ADCs are fast compared with, say, the frequency of the mains
supply (50 or 60 Hz), so that mains frequency (or double mains
frequency) interference and ripple on rectified signals can cause a
problem. Figure 6.3 illustrates the problem; the interface under the
control of the micro samples the analog input signal during a
relatively short period of time, and the analog signal has a
significant mains frequency component which results in widely
fluctuating values of the digital input. The level of interference
shown in fig 6.3 is not unusual on a signal intended for a chart
recorder (indeed this author has purchased more than one brand of
chart recorder which actually injected this amount of mains frequency
ripple onto a 10 mV input signal). The problem can be usually overcome
by filtering the analog signal with a simple RC network having a
corner frequency of, say, 10 Hz - if such action will not mask the
expected rate of signal variation. Alternatively, impedance conversion
close to the signal source may be used to avoid interference pick-up
on the analog input line.

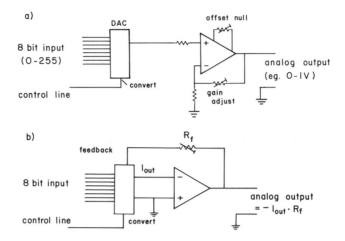

Fig 6.4   Typical analog output interfaces. a) shows a circuit based on
           a voltage output DAC, while in b) a current output DAC is
           followed by a transresistance amplifier to generate a voltage
           output.

6.1.2 Analog output interface

      An analog output (or digital-to-analog conversion) interface
performs the opposite translation, taking a byte value between 0 and
255 and converting it to an analog output signal. Two representative
analog output interfaces are shown schematically in fig 6.4, where a)
a voltage output DAC (such as the ZN428E) and b) a current output DAC
(such as the AD7523) are illustrated. Again conversion from the binary
value present on the 8 parallel data lines may be initiated by the
application of an appropriate control signal level to the conversion
enable pin of the DAC IC, although not all DAC circuits provide this
facility, and many offer a latchable buffer to hold the digital data
so that an analog output may be maintained after a digital input
signal has been removed.

      Voltage output DAC circuits produce an output voltage typically
in the range 0-1 V to 0-10 V, although most have relatively high
impedance outputs which, to avoid temperature fluctuations, require
buffering if a high accuracy output is required. The output voltage
may be scaled by a selectable gain amplifier or attenuator to lie in
any desirable voltage range (as illustrated in fig 6.4a), or may be
passed to a transconductance amplifier to generate a required output
current. Multiplying DACs allow a wider choice of output ranges, and
the reference signal (which determines the output range) could be
supplied be a second analog output circuit, allowing computer control
of the range. Current output DACs produce an output which is typically
in the range 0-1 mA, and which may be converted to an output voltage

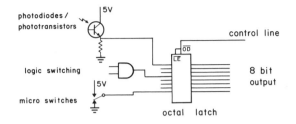

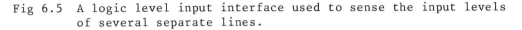

Fig 6.5 A logic level input interface used to sense the input levels
         of several separate lines.

using a transresistance amplifier. Many current output DACs provide an
on-chip feedback resistance for use with a transresistance amplifier,
so that only a good quality op-amp is necessary to complete a voltage
output circuit. An additional and variable feedback resistance may be
incorporated (as in fig 6.4b) if a variable conversion gain is
required.

DAC ICs are usually faster than comparably priced ADC chips and
for most laboratory applications the speed of conversion does not form
a limitation (see section 4.7.1). Analog output interfaces are
frequently used to provide voltage signals for pen recorders and
control signals for variable power devices such as lamps or heaters.
They can of course be used to provide on/off signal levels, although
generally the switched output interface would be more appropriate for
this task.

6.1.3 Logic input interface

The logic input interface illustrated in fig 6.5 uses an octal
latch to detect logic levels on up to eight sense lines. The signals
in this case normally arise from relays, microswitches or logic
circuits, although in some cases it is possible to use variable analog
signal sources - provided that one remembers that the computer will
only sense whether the voltage level is hi or lo (ie >2.5 V or <2.5 V
for a CMOS latch operating at 5 V. TTL is not a good choice in this
application unless there is no doubt about the ability of the source
to sink the TTL lo level current). However, if there is a risk that a
voltage level being sensed may lie in an undefined region (eg at 2.5 V
for a CMOS system or in the range 0.8 - 2 V for TTL) then it would be
better to convert the voltage to a logic level using a comparator with
an appropriate comparison (threshold) voltage (as described in section
4.6.1).

In operation the hi/lo level of each of the eight sense lines is
stored in the buffer as one bit of an 8-bit byte when the appropriate
control signal level is applied to the latch enable pin of the IC. The

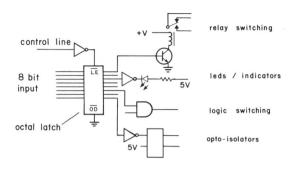

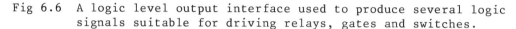

Fig 6.6   A logic level output interface used to produce several logic
          signals suitable for driving relays, gates and switches.

byte is then read by the computer as a binary number, such as
10010111, where a 1 indicates sense level hi and a 0 indicates sense
level lo. Thus a value of 255 would indicate that all sense lines were
hi. Once the byte has been read by the computer the individual bit
values may be decoded by software using the techniques described in
section 5.7. The byte value represents the states of the sense lines
at the time the latch enable signal is received by the buffer, a time
which may be selected with considerable precision (ie. a specific
microsecond) but may be earlier than when byte is read by the computer
- particularly if the operation of the interface is being controlled
from a BASIC program (see section 5.7). Typical applications are in
safety devices (eg. ensuring that doors are closed before X-ray tube
power supplies are activated, cf fig 4.4), sensing the presence or
absence of a sample on a conveyor belt, and reading a manually
switched control panel.

6.1.4 Logic output interface

     A complementary type of interface is the logic level output or
switched output interface (fig 6.6) which produces eight output logic
levels of 0s and 1s, ie. nominal voltages of 0 and 5 V (of course,
these may be readily converted to other voltages or currents). In this
case a byte from the computer is stored in the octal latch when its
latch enable pin is activated, and each line of the buffer output
produces either 0 V if the corresponding bit is a 0, or 5 V if the
corresponding bit is a 1. The latch used may be CMOS or TTL and,
although the latch shown is a TRI-STATE variety (eg. 74C373 or
74LS373) with a grounded output disable pin, there is no reason why a
non-TRI-STATE device should not be used. If only momentary logic level
outputs are required, the output disable pin may be connected to the
latch enable pin - in which case outputs will only be present when the
control line is lo. This technique is useful when one wishes to
operate an increase/leave unchanged type of control.

While these output signals may be used directly (eg. for operating low current indicators such as LEDs) it is usually better to buffer the output signals with transistors, TTL drivers or analog buffers such as voltage followers, if the load is likely to exceed a few mA, or to use logic activated, solid state relays or opto-isolator switches if it is necessary to switch high currents or voltages or ac. Opto-isolator switches are switches whose operation is controlled by the light produced by an LED housed in the same package as the switch. Opto-isolator switch ICs are particularly valuable for interfacing as they allow reliable electrical isolation between units attached to the computer and units attached to other powered systems. Typically potential differences of thousands of volts between the two circuits can be tolerated. In addition to simple switching applications, modern opto-isolator devices are available containing light-operated high gain transistors (Darlingtons), silicon controlled rectifiers (SCRs) and triacs. The latter are ideal for providing isolated logic level to mains interfaces. Switched outputs find application in automatic sampling and injection, in operating indicators and warning signals, in turning on and off controlled elements such as heaters or light sources, and in control of multiplexors and selectable gain amplifiers (see below).

6.1.5 Digital data interfaces

Both the logic input and logic output interfaces can, of course, be operated as 8-bit digital transmitters, and in this guise find application for communication with instruments which have facilities for 8-bit data transfer. These interfaces are often used with opto-isolator buffers, to minimise any risk of a fault in one system causing damage in another. The use of opto-isolator buffers also enables non-TTL logic levels to be interfaced to a computer with minimal difficulty. Many common laboratory instruments which have digital displays (eg. pH meters, frequency counters and digital multimeters) also provide multidigit BCD outputs. Typically these may take the form of 3.5 digits (ie. 0-1999) and a sign "bit" formed into 14 parallel digital outputs, along with a hi/lo "data valid" signal. This type of data can be transferred using a 16 bit digital input interface, such as that described in section 6.3 and fig 6.14. Of course if BCD or character data is transferred to a computer, then the user's program must be able to use or decode the bytes as required; thus the same byte values read from a BCD source and from a binary source represent different quantities (see section 5.1). 8-bit digital output interfaces are also useful when several different interfaces are connected to the same computer, and this subject will be discussed further in section 6.2.

6.1.6 Pulse counter interface

A pulse counter interface accepts pulses (usually logic pulses or pulses with specified characteristics which are converted into logic

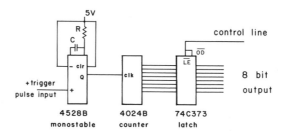

Fig 6.7   A typical pulse counting interface, which may also be used
          for timing applications by counting pulses from a clock
          oscillator.

pulses) and counts them in a binary counter (see section 4.5). A
typical pulse counter interface is shown in fig 6.7. The input pulses
are converted into fixed width logic pulses using a 4528B monostable,
in this case triggered by the positive going edge of the input pulse.
The width of the output pulse is chosen to suit the counter circuit,
and in this case could be set to 1 microsecond by the choice of R=10 k
and C=22 pF. The counter is an 8 bit binary ripple counter (4024B),
which can only count from 0 to 255, and it is important to realise
that pulse number 256 returns the byte value within the counter to
zero, so that the user's program needs to check for this "byte
overflow" and correct the values in the program accordingly (typically
by adding 256 to a reading which is smaller than the previous
reading). Furthermore the byte stored in the buffer represents the
count at the time the latch enable signal is received by the buffer
and will not change if further pulses are counted before the byte is
transferred to the computer.

     Pulse counter interfaces are of particular value in systems
involving X-rays, radioactivity, photon counting or precision timing,
although 16 or 24 bit interfaces are likely to be of more value than
8-bits when high pulse rates are involved (see section 6.3). Pulse
rates or frequency may be determined using pulse counters by, not
surprisingly, determining the count within a specified period of time.
Time intervals can be recorded by using a pulse counter to count the
pulses from a crystal oscillator (see section 6.4), and this is
sometimes a more reliable technique than trying to use the
microcomputer's own BASIC "clock" (if it has one), particularly on
micros where the processor is interrupted frequently (for example, the
PET and Apple 6502 processors are interrupted at the mains frequency).
Of course, the MPU's clocking signal may be used as a source of
pulses, but there is no reason to suppose that this will provide a
high stability source of exactly the expected frequency. Furthermore,
the signal used should be the control bus version, rather than the
MPU's clock input, as most post-1976 MPUs have their clock circuits

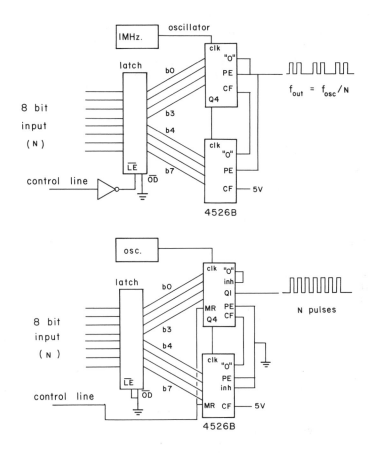

Fig 6.8   Two examples of pulse output interfaces. a) generates logic
          pulses at a specified frequency, b) generates a specified
          number of pulses.

on-chip (only the crystal is externally connected). Some micros cannot
tolerate capacitive loading of the clocking signal, so that the line
requires buffering within the micro and before lengths of external
cabling are attached.

6.1.7 Pulse output interface

     Pulse output interfaces can also be constructed, although where
pulses are used in instrumentation (eg. for stepping motors or clocked
logic control) the height, width and rise time of each pulse is likely
to be important, so that the interface in this instance is often
supplied as an integral part of an instrument or sold for the control
of a specific device (such as the Bentham stepping motor interface).
Furthermore, two distinct categories of pulse output interfaces can be

distinguished: those for which the output consists of a continuous
stream of pulses or a square wave at a selected pulse rate; and those
for which the output consists of a selected number of pulses. The
former variety is useful where a particular frequency signal needs to
be applied to a sample, or where a "chopping" signal is required. The
latter category has applications in dispensing discrete samples of
material and in providing pulse trains for the operation of stepping
motors.

Typical circuits representing each type of pulse output interface
are shown in fig 6.8. Both are based on CMOS 4526B programmable
divide-by-N counters driven by a 1 MHz crystal oscillator, and both
may have their outputs buffered to provide TTL compatible pulses. The
4526B is a 4 bit device, but is designed so that several devices may
be cascaded. Its "0" output (that is what it's called) is normally 0,
but can change to a 1 when the count reaches 0000 (ie. when it
overflows) if the level on the cascade feedback (CF) pin is a 1. Thus
in fig 6.8a the high order nybble device has its CF pin connected to
the 5 V supply, allowing the "0" output to change from 0 to 1 when the
high order count reaches 0000 (it starts at the value set by b4-b7).
This output is connected to the CF pin of the low order nybble device,
so the "0" output of the low order device is unable to change until
after the high order count has reached 0000 and the low order device
subsequently overflows. The preset enable (PE) pin of each device is
connected to the low order nybble's "0" output, so each time this goes
high (which occurs each time the number of clock pulses specified by
the data byte have been counted), the two devices reload their
respective nybbles from b0-b7 and start counting again (which returns
the circuit's output to 0). So the circuit in fig 6.8a produces a
continuous stream of output pulses at an average frequency equal to
the clock frequency divided by N, the value of the data byte. (Note
that this technique does not produce equally spaced pulses for all
values of N.)

In fig 6.8b the circuit's output is taken from the Q1 output of
the low order counter, and this output changes level each time a clock
pulse is counted - until that is, the preset count is reached and the
device's "0" output goes hi and inhibits all the Q outputs by holding
the inhibit pin hi. This can only occur after the high order device
has counted its preset count, produced a hi output on its "0" output,
and taken the low order device's CF pin hi. Thus the circuit produces
the number of clock pulses preset by the data byte, and then does
nothing until reset by a logic 1 on the master reset (MR) pin. Return
of the MR pin to 0 allows another pulse train output of the number of
pulses specified by the data byte.

6.2 Multiplexing

Each of the interfaces described above may be connected directly
to an 8 bit port of a micro when a single control line is available.

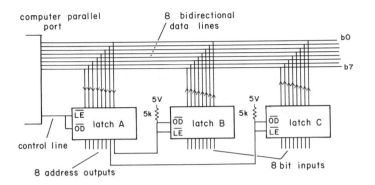

Fig 6.9   A part circuit for addressing one of several interfaces
          connected to a bus. This system does result in bus
          contention, so the latches should be CMOS (eg. 74C373). Ways
          of avoiding the contention are shown in fig 6.12.

Thus unless the micro is fitted with several ports (as could be the
Apple, with its 8 I/O slots), we are limited to a single interface
between the micro and any laboratory instrumentation. For most
purposes this would be unsatisfactory. However we can overcome this
limitation by using the computer's parallel port as a bus, ie.
connecting several different 8 bit circuits in parallel with the port.
Naturally this requires a certain amount of care, as we must ensure
that no more than one circuit attempts to set the digital signal
levels on this bus - so that our circuit devices which can determine
these signal levels must have TRI-STATE outputs.

     One of the simplest techniques for the implementation of a bus
can be seen in the part circuit shown in fig 6.9, (we will come to the
missing items later). First of all we note that, in addition to the
data lines of the computer port, three octal latches are connected to
the bus, one latch (A) is connected at its input lines and the other
two at their outputs. The latch enable and output disable pins of
latch A are both connected to the computer port's control line, and we
shall assume that initially this line is hi. The latch enable and
output disable pins of the other latches are held hi by pull up
resistors (so the outputs of these latches are normally held in the
high impedance state). Thus the computer is able to output a byte onto
the bus without either B or C attempting to do the same.

     Let us imagine that the latches B and C are holding 8 bit data
which we wish to read into the computer via the bus. A typical
sequence for carrying out the data transfers is as follows:

1. The computer outputs a byte with a binary value, eg. 254 (ie.
   11111110).

2. The computer forces its control line lo to operate the latch enabl and output disable functions of latch A. This results in 254 being latched into A and output on the output lines which carry individual bit levels to the output disable pins of the other latches.
3. The line from latch A (d0) connected to the output disable pin of latch B now carries a 0 (ie. lo level), so this latch latches the state of its input lines and outputs the byte value onto the bus. The other latches have their output disable pins held hi (d1 - d7 of latch A), so their outputs remain in the high impedance state. (There is actually bus contention at this point - both the computer and latch B are trying to hold bytes on the bus. For this technique to work the computer user port needs to be buffered with fairly powerful TTL bus drivers, while the latches in fig 6.9 should be CMOS. Under these conditions the computer wins without any damage being done, although the clash is short-lived anyway. We have used this technique for several years with PET computers and experienced no trouble, but it could damage a computer fitted with feeble port drivers. An alternative arrangement which avoids this difficulty altogether is discussed below.)
4. The computer stops the output of 254 and reconfigures its parallel port for input. The data on the bus is now the byte value being output from latch B.
5. The computer inputs the byte value on the bus lines (eg. X = PEEK(59471) in the case of the PET user port).
6. The computer sets the control line hi, disabling the output of latch A, so that this latch is no longer holding the output disable lines of any latches. The output of the other latches become or remain disabled by the effects of the pull up resistors connected to their output disable pins, so latch B is no longer asserting a byte value on the bus and the bus is free again.

Repeating the above sequence with a computer output of 253 (ie. 11111101) would allow the byte held in latch C to be read over the bus. As there are eight output lines from latch A we could easily arrange to read the contents of eight latches using this technique, each latch being addressed with a byte consisting of one 0 and seven 1s (corresponding to values of 254, 253, 251, 247, 239, 223, 191 and 127).

A somewhat more efficient way of decoding the addresses for latches connected to a bus can be based on a decoder IC. This approach is illustrated in fig 6.10, where the decoder the CMOS 4515B device called a 1-of-16 decoder or demultiplexor. This particular IC utilises a 4 bit input and in the circuit diagram the input pins are connected to the low order lines (ie. b0 - b3) of the bus. The IC has 16 output pins, S0 - S15, and normally 15 of these are hi and 1 is lo, the lo one being selected by the binary value of the 4 bit (0 - 15) input. Thus an input of 0 (0000) selects the output on pin 11 (S0) to be lo, and the other 15 outputs to be hi; an input of 5 (0101) selects a lo

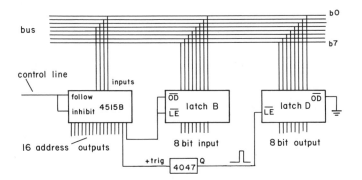

Fig 6.10 Using a 4515B demultiplexor to address 1 of 16 single byte
         latches connected to a bus. Only the lower 4 bits of the bus
         are needed for addressing via the 4515B.

output on pin 6 (S5), and an input of 15 (1111) selects a lo output on
pin 15 (S15). The device is not a TRI-STATE device, so it does not
have an output disable facility of the type present in the octal latch
we used above. However, the 4515B does have an output inhibit pin, and
when the level applied to this pin is hi all the output pins are
forced hi (level, not high impedance) irrespective of the digital
input. Because the output is always hi when it is not specifically
selected to be lo, the pull up resistors used in fig 6.9 are not
required when this decoder is used. There is also a "follow" pin
(analogous to the latch enable connection of the octal latch). When
the follow pin is hi the value decoded follows the coding of the 4 bit
input, but when follow is made lo the value of the 4 bit code present
at the digital input is stored internally and used for decoding until
such time as follow returns hi again. Connecting the follow and
inhibit pins together allows the circuit to be controlled by a single
hi/lo control level from the computer.

    If one considers the sequence of steps described for fig 6.9
while examining the interconnections of fig 6.10, then it will be seen
that the circuits achieve the same results - but in the case of fig
6.10 we may select the contents of up to 16 different latches for
transfer over the bus. Larger number of addresses can be used, for
example, by using a second decoder with input taken from b4 - b7 of
the bus, although in its simplest implementation this would require
that one address for each decoder (such as 0000) was left unused so
that the two decoders did not enable two different latches at the same
time. Both of these address selection techniques produce the same end
result and there are numerous other devices which could be used to
perform this function. In many of the subsequent figures we shall use
selectors based on inhibitable decoders (eg. 4515B) and TRI-STATE
latches (eg. 74C373) more or less interchangably. Of course when a

TRI-STATE device is used in this particular role pull-up resistors are
needed, although to simplify circuit diagrams these have been omitted
in the remainder of this chapter.

An output latch D has been included in the circuit of fig 6.10 to
illustrate one way in which an output can be obtained continuously
(eg. for a chart recorder), while its 8 bit value may be changed at
will. The output disable pin of this latch is grounded, so the device
is always exerting an 8 bit output. However, the latch enable pin is
connected to the output, Q, of a 4047 monostable, and as this level is
normally lo the data present at the input pins of the latch is not
normally passed through to the outputs — any previously latched data
byte forms the output under these conditions. When the relevant
decoder output (line 1 in this example) is selected it goes lo, and
the required output byte may be loaded onto the data bus by the
computer, although at this stage the outputs of latch D are not
changed. When the decoder outputs are inhibited line 1 goes hi again.
This lo-to-hi transition triggers the monostable which produces a lo-
hi-lo pulse on its output and this allows latch D to latch the data
currently present at its inputs. As the OD line is still grounded the
8 bit output changes immediately to the newly latched value, and stays
like that until the latch is enabled again by the decoder/monostable
combination.

A typical BASIC subroutine to output a byte through latch D via
the PET user port is as follows:

```
100 POKE 59469,255: REM sets DDR for output
110 POKE 59471,1: REM output address 1
120 POKE 59468,PEEK(59468) AND 31 OR 192: REM control lo
130 POKE 59471,X: REM output byte X
140 POKE 59468,PEEK(59468) OR 224: REM return control hi.
```

Similarly a subroutine for byte input is:

```
200 POKE 59459,255: REM set DDR for output
210 POKE 59471,0: REM output address 0
220 POKE 59468, PEEK(59468) AND 31 OR 192: REM control lo
230 POKE 59459,0: REM set DDR for input
240 X=PEEK(59471): REM read port data bus
250 POKE 59468, PEEK(59468) OR 224: REM return control hi
```

The instructions used for control line toggling are described in
section 5.7.

A technique capable of an even wider range of addressing is
illustrated in fig 6.11. One decoder connected to the high lines of
the bus (b4 - b7) is used to select one of several (up to 16)
decoders, each of which selects a latch from 1 of 16 using a code from
the low order lines of the bus (b0 - b3). This allows up to 256

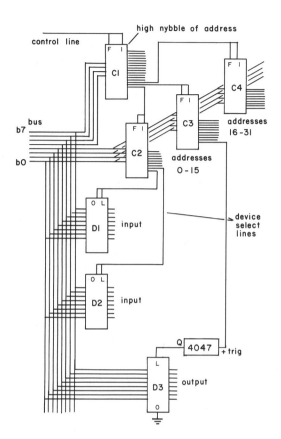

Fig 6.11 An interface addressing system capable of extension to 256
          addressed lines. The Ci are 4515B 1-of-16 decoders, while the
          Di may be 74C373 latches.

different digital signals to be selected, which should be enough for
even complex systems.

     All of the above circuits produce a momentary clash on the data
lines when any input interface is addressed. While as stated above
this technique works perfectly well for any TTL buffered PIA port when
CMOS TRI-STATE latches are used for the bus connections, there is a
straightforward way of avoiding the clash — although at the cost of an
additional control line operation and a consequent reduction in the
speed of operation. The technique is illustrated by the part circuit
shown in fig 6.12a. The control line is connected to a 4013B flip-flop
(a circuit which can be arranged to change its output level for every
hi-to-lo transition on its clock input), and to two OR gates. The
control line is toggled hi-lo-hi with the required device address
present on the data lines, and this results in a latch enable pulse

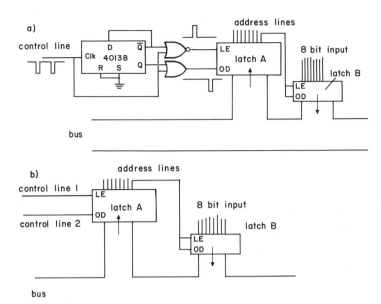

Fig 6.12 Two ways of avoiding bus contention with an multiplexed
         interface system. a) The use of two togglings of a single
         control line to operate the address latch, A, by alternate
         control of its latch enable and output disable lines. b) The
         use of two independent control lines to operate the latch and
         output disable lines of latch A.

being passed to latch A and the device address being latched. However,
the outputs of latch A remain disabled until a second hi-lo transition
occurs on the control line, and this is carried out only after the PIA
port has been reconfigured for input. Once this occurs the addressed
latch (eg. B) deposits its data onto the data lines where they can be
read by the computer, and the PIA control line returned hi to
terminate the transaction. A typical BASIC subroutine for the input of
a single byte is as follows:

```
200 POKE 59459,255: REM set DDR for output
210 POKE 59471,0: REM output address 0
220 POKE 59468, PEEK(59468) AND 31 OR 192: REM control lo
225 POKE 59468, PEEK(59468) OR 224: REM control hi
230 POKE 59459,0: REM set DDR for input
235 POKE 59468, PEEK(59468) AND 31 OR 192: REM control lo
240 X=PEEK(59471): REM read port data bus
250 POKE 59468, PEEK(59468) OR 224: REM control hi
```

The corresponding control level changes are included in fig
6.12a. A similar arrangement can cater for byte output without the use

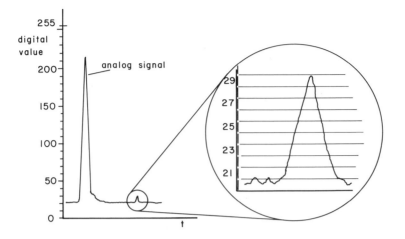

Fig 6.13 The limitation of a low resolution analog interface is
         illustrated by the conversion of the smaller analog signal
         peak within a small digital range.

of the monostable pulse circuits of figs 6.10 and 11. And of course
the technique can be adapted for use with address decoders rather than
address latches by appropriate use of the follow and inhibit pins. An
alternative technique may be implemented by operating the address
latch's latch enable and output disable pins with separate control
signals from the computer as illustrated in fig 6.12b. This is a
simple way of avoiding bus contention, but of course is limited to
those micros on which a second output control line may be easily used.
In the case of the PET a second control line is available is the
cassette ports are not in use, as the cassette write line has its
level determined by the content of the VIA address 59456 (b3) and is
available on pin 7 of the user port.

6.3 Multiple byte interfaces

     An 8-bit interface uses data bytes with values between 0 and 255,
and the smallest possible difference between byte values is 1 (eg. 122
and 123). An important consequence of this is that the resolution of
the interface is only 1 in 255, equivalent to about 0.5% of the full
scale value. While this may be acceptable in some instances (eg.
counting infrequent events), it is easy to see that there are many
cases in which the resolution of an 8-bit interface is inadequate.
Figure 6.13 shows a chart record of an analog signal from a
chromatograph, with the byte values produced by an 8-bit ADC marked on
the deflection scale. While the larger peak covers most of the 0-255
scale, and is therefore converted with reasonable precision, the
smaller peak would be converted within the range 20-30 so that the
peak height estimated from byte values would contain an uncertainty of

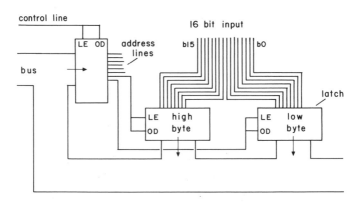

Fig 6.14 A 16 bit digital input interface, which transfers its data
        over the 8 bit bus as two separate bytes controlled by two
        address lines.

around 10%. Probably the most common solution to this problem is to
use a higher resolution interface, ie. an interface which converts the
instrumental signal into 12, 16 or 24 bits.

    A 16-bit resolution ADC interface converts an analog input signal
into a 16-bit binary number (with a value between 0 and 65535). As the
computer can accept only one 8-bit byte at a time obviously the 16-bit
data must be transferred to the computer as two separate bytes. This
can be accomplished by storing the 16-bit data in two 8-bit buffers
whose outputs are connected to a parallel bus as shown in fig 6.14.
One buffer now deals with bits 0-7 of the 16-bit data, and the other
with bits 8-15. This arrangement allows us to transfer bytes via the
bus in much the same way as we did with the multiplexed 8 bit latches
in section 6.2. Some BASIC instructions to transfer the two bytes and
combine them to form the 16-bit value are shown below:

```
200 POKE 59459,255: REM set DDR for output
210 POKE 59471,0: REM output high byte address 0
220 POKE 59468, PEEK(59468) AND 31 OR 192: REM control lo
230 POKE 59459,0: REM set DDR for input
240 X=PEEK(59471): REM read high byte from data bus
250 POKE 59468, PEEK(59468) OR 224: REM control hi
260 POKE 59459,255: REM set DDR for output
270 POKE 59471,1: REM output low byte address 1
280 POKE 59468, PEEK(59468) AND 31 OR 192: REM control lo
290 POKE 59459,0: REM set DDR for input
300 Y=PEEK(59471): REM read low byte from data bus
310 POKE 59468, PEEK(59468) OR 224: REM control hi
320 Z = X*256 + Y: REM Z contains 16 bit value.
```

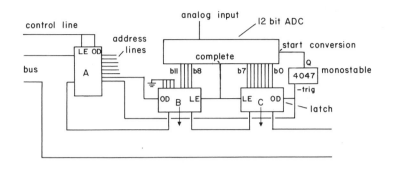

Fig 6.15 Using an interface address to control the action of an
interface system. In this case the ADC's "start conversion"
line is pulsed (via the 4047) when the low byte is addressed,
and both latches are latched by the ADC's "conversion
complete" signal. The ADC used is the successive
approximation ADC1210; it requires additional circuitry (a
clock) not shown.

Note that the values stored in both X and Y must be in the range 0-
255, while the 16-bit value produced in line 320 and stored in Z may
be 0-65535.

Most of the types of interfaces described earlier are readily
available in 12 or 16 bit versions, and 24 or 32 bit versions can be
obtained or constructed. For analog conversion interfaces 12 bits
allows a resolution of 1 in 4095, 16 bits 1 in 65535, and 24 bits
about 1 in 16 million. However, increasing the resolution of ADC and
DAC interfaces does dramatically increase the cost of the interface,
particularly if the retention of high speed conversion is necessary.
The increased number of program steps required for multibyte data
transfers also results in the speed limitations of BASIC soon becoming
apparent. While up to 10 single data byte transfers per second may be
achieved using BASIC with simple 8-bit interfaces in the manner
illustrated above, it becomes difficult to reach 2 readings a second
when 24-bit interfaces are used. So, unless the interface supplier
also provides machine code software for data transfers, the user may
have to write his own, or use a BASIC compiler, in order to make his
program run at a useful speed.

6.4 Interface control

Interfaces with resolution greater than 8 bits and interfaces
which are required to provide several quantities recorded at the same
time, require more precise control of their fuctions than can be
achieved with the circuits we have considered so far. For example, if
data is to be read into a computer from a 12 bit ADC interface, then

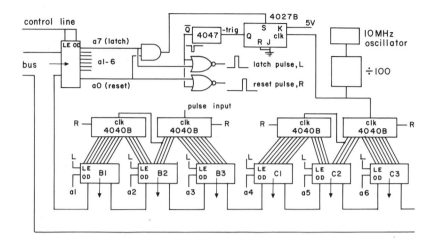

Fig 6.16 A 24 bit pulse counter and 24 bit timer which may be
zeroised, latched and read using addresses from the
computer.

the 12 bit data must not change between the transfer of the first 8
and last 4 bits. One way in which this can be accomplished is
illustrated in fig 6.15, where the ADC's "start conversion" signal is
supplied by a monostable (generating, say, a 1 microsecond pulse,
although the timing components are not shown in the figure) triggered
by the same select line as that which enables the output of latch C.
The contents of both latches are latched by the ADC's "conversion
complete" signal, which is assumed to be a lo output or a lo-hi-lo
pulse (obviously an inverter could be included if the end of
conversion was signalled by a complementary signal). Thus 8 bits may
be read from latch C, then latch B selected and four bits read from
this without the ADC's 12 bit data changing. For this circuit it would
be necessary to ensure that the time delay between selection of latch
C and the reading of the byte by the computer was longer than the
conversion time of the ADC - not usually a problem when the
controlling program is written in BASIC. Of course, it is also
essential that latch C is addressed before latch B!

A more versatile approach involves dedicating lines from the
address decoder for each function required by the interface. A fairly
detailed example of this technique is included, both to illustrate the
high degree of control which may be achieved, and to enable us to
highlight a few aspects of timing accuracy. The circuit in fig 6.16
for a multifunction interface consisting of two 24 bit counters, one
which counts pulses from a measuring transducer (such as a
photomultiplier - signal conversion circuits are not shown), and the
other which counts pulses derived from a 10 MHz crystal oscillator,
but divided down by 100 (a variable divider was used in the original

to increase the useable timing range). This unit forms part of a
system designed for the measurement of photon count rates
(proportional to detected light intensity), and the clock was included
so that high precision measurements could be made.

Let us consider first the decoder, which in fig 6.16 is a 4515B
1-of-16 decoder. Eight of its output lines are used (the original
circuit had several additional functions), 6 providing latch select
signals (a1 to a6) to enable the latches to deposit their data bytes
on the bus, one providing a latch signal and one providing a reset
signal. The latch signal was required to operate all six latches at
once, enabling each to latch the appropriate byte from the relevant
counter. Similarly the reset signal was required to reset all four (12
bit) counters to zero, so that both the photon counter and the timer
were simultaneously restarted from zero.

The latch and reset signal could be used directly by connecting
these conductors to the appropriate pins on the latches and counters.
However, this would result in the time intervals recorded containing
an uncertainty of up to two oscillation periods (one for the reset at
the begining of the time period, and one for the latch at the end).
This would have been acceptable for timing periods of the order of
2^{20} cycles (ie. a few hundred seconds) but was unacceptable for
periods of less than a second (2^{10} cycles), for which the error was
about 0.1%. A convenient way of improving the timing precision may be
implemented by applying latch and reset pulses to the latches and
counters only on the falling edge of the clock's waveform, which can
be done by using the computer generated signals to enable a JK flip-
flop, whose output only changes when the edge of the clock pulse
arrives at its clock input (see section 4.6.2). The output's falling
edge is then used to trigger the operation of a monostable which
produces an output pulse of 1 microsecond duration, which in turn is
gated by one of the two NAND gates and applied to either the latch or
the reset pins of the latches or counters respectively. In this way
the time period for which the counters are operational may be recorded
in units of 0.01ms, but with an accuracy and repeatability of better
than a few tenths of a microsecond.

Thus when a reset address is issued by the computer, the
normally-hi S input of the JK flip-flop circuit drops to a 0. The
flip-flop then has J=0, K=1 and Q=1, and waits for the next falling
edge of the clock waveform before its output can fall. This transition
then initiates the output of a 1 microsecond wide pulse from the
monostable to reset all of the counters to zero. Counting proceeds.
When the computer needs to read the counter and timer, a latch address
is issued and, after the delay required to ensure synchronisation with
the clocking oscillator, the 6 bytes of data from the counter and
timer are latched into the six latches. These bytes are then read by
the computer one at a time in response to the appropriate address
codes. If the computer is satisfied with the readings the counters may

be reset to zero for a second measurement, or alternatively (as the
counter and timer were left running even when the latching operation
took place) a second latch address may be issued and the updated byte
transferred to the computer. This technique is particularly valuable
for the counting of randomn pulses as it enables the data to be
examined repeatedly until a required statistical error limit has been
reached.

6.5 Handshaking

While the interfaces described in section 6.1 are all very
straightforward, there are circumstances in which communication
between any one of them and the microcomputer could be unreliable. The
main problem lies in the use of a single control line to activate the
interface. Using one control line operated by the computer results in
the computer functioning blind – it has no way of knowing whether the
interface is ready for a byte transfer. In most cases the system will
work correctly under the control of a BASIC program, because BASIC is
relatively slow and several milliseconds elapse between each operation
connected with the interface. However, even this may not be enough
time for a high resolution ADC to complete a conversion, or for a
pulse output interface to discharge a specified number of pulses
before receiving another instruction. The situation becomes even more
serious if a compiled language or machine code is used for interface
control and byte transfer operations, and a number of time delay loops
may be required in the controlling program to prevent consecutive
signals being sent to the interface too quickly.

Some devices connected via an interface to a computer may only be
able to handle data bytes much more slowly than the computer can
despatch them, even when under the control of a BASIC program. Low
cost, unbuffered printers are an obvious example, being able to print
at perhaps 80 characters per second while the computer can transmit
several hundreds or thousands of bytes during the same period.

A simple solution to this problem may be found in using two
control lines, one carrying a signal from the computer to the
interface (as before), and a second carrying a signal in the other
direction. The basic principle is illustrated in fig 6.17, where the
Cx1 and Cx2 control lines of a typical PIA port are used to carry
these "handshaking" signals to a simple (ie. non-multiplexed)
interface. We will discuss the byte transfer process used for output
of the data byte first, because many micros are equipped with an
output port which utilises these principles, often for connection to a
printer. (The Centronics[R] parallel interface is closely related to
this system but requires open collector data bus drivers.) The signal
levels used during the transfer are included in fig 6.17.

Firstly we should note that the PIA is programmed so that it
responds to a lo-to-hi transition on its Cx1 control line by setting

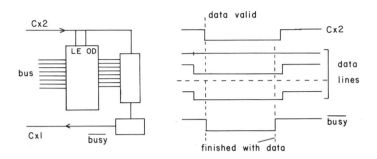

Fig 6.17 A two wire "handshaking on output" technique. The output
interface (left) generates a busy signal until it has
finished with the data. The control line signal levels are
illustrated on the right.

the PIA interrupt flag. Of course, it would be equally valid to
program the PIA so that it responded to a hi-to-lo transition if the
interface hardware gave signals complementary to those illustrated in
fig 6.17. When the computer wishes to "handshake" a data byte to the
interface the following sequence occurs:

1. The computer places a byte on the port data lines,
2. The computer holds Cx2 lo to signify that the data on the data
   lines is valid, and then waits for the PIA interrupt flag to be
   set.
3. The interface circuitry holds the Cx1 line lo, to indicate that it
   is busy digesting the data on the data lines.
4. When the interface has digested the data byte it returns the Cx1
   line hi, which sets the interrupt flag in the computer's PIA.
5. The computer observes that the interrupt flag has been set. It
   clears the flag and returns Cx2 hi (to indicate that there is no
   valid data on the data lines).
6. If any more data are to be sent the cycle starts again at step 1.

   For byte input the situation is slightly different as it is the
interface or peripheral which must place the data byte on the data bus
and signal to the computer that the data bus is ready for reading. An
example of data input is given in fig 6.18 for an analog input
interface handshaking data bytes to the computer with a conversion
sequence before each byte is ready for transfer. In this example the
following sequence occurs:

1. The computer drops the Cx2 line from its normally hi level to lo,
   which initiates a conversion cycle in the ADC, and waits for the
   PIA interrupt flag to be set.
2. The ADC sets the Cx1 line lo to indicate that conversion is in
   progress and the data bus is not ready. (If the ADC's busy signal

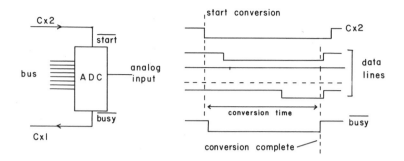

Fig 6.18 A two wire "handshaking on input" technique, illustrated wit
        an 8 bit ADC which generates a busy signal until conversion
        is complete and the 8 bit output is stable on the data line.
        The control line signal levels are illustrated on the right

operates the other way around, an inverter may be included in the
Cxl line. ADCs which generate a conversion complete pulse will
work whatever the sense of the pulse.)
3. When the conversion cycle is complete and the digital output is
   loaded on the data lines and stable, Cxl is returned hi by the
   ADC. This sets the interrupt flag in the computer.
4. The computer finds that the PIA interrupt flag has been set. The
   computer then clears the flag, reads the data bus and returns the
   Cx2 line hi.
5. If another transfer is required the cycle starts again at step 1

      The techniques for byte input and output with handshaking are
very reliable and moderately fast (typically up to 1000 bytes a secor
may be transferred under the control of a machine code program) if the
interface imposes no significant limitation. Any of the interfaces
described earlier can be adapted for use in a handshaking mode,
although if one is forced to use a single port for both interface
addressing and data transfer the system can become a little
cumbersome. (Of course, addressing, as used in section 6.4, involves
output through the port. But as this only requires the operation of a
latch at the interface end there is little to be gained in terms of
speed from handshaking the address byte.)

      An example of an addressed input interface system is given in fi
6.19. The address is transmitted from the computer and is latched int
the address decoder using the Cx2 control line. The computer then
waits for the interrupt flag to be set by the interface system to
indicate that a byte is ready for reading. For some devices a
hardwired busy signal (generated by the monostable which is triggerec
when Cx2 goes lo) of a couple of microseconds duration is adequate.

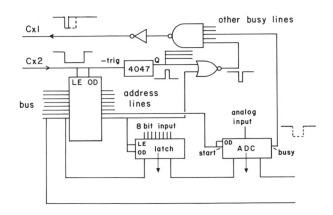

Fig 6.19 An addressed interface system providing handshaking on input
to enable byte transfers to occur as fast as possible.

This signal can be ANDed with the latch enable/output disable signal
of the latches used for pulse counter, digital and logic input
interfaces, as these will operate fast enough to get the data onto the
bus in time for the computer to read it. A slower device, like the ADC
circuit (which generates its own busy signal) can be connected, via
the multiple input NAND gate and the inverter, to the Cxl line and not
make use of the simulated busy pulse generated by the monostable.

6.6 Synchronous byte transfers

    The byte transfer techniques we have considered so far have been
asynchronous techniques, ie. addressing the interface and transferring
a data byte have been operations independent of the MPU's clocking
signal. This has required that the interface address and the data byte
are transmitted over the port bus one after the other, and that
control signal and data direction registers in the PIA are changed
during each transfer operation. Consequently none of the techniques
described qualifies as "fast", even though the use of machine code may
allow transfer rates of a thousand bytes per second or so. However,
there is another method of communication between the computer and
external devices, which operates in much the same way as byte
transfers between the MPU and ROM or RAM, and allows us to make use of
the speed characteristic of those transfers.

    As the MPU's 16 bit address bus is generally available for use
outside the computer (even if partially decoded), we can use this to
carry addressing signals to an interface unit. Furthermore the timing
signals available on the control bus (see section 5.8) provide the
means of operating latches for the latching of address bus codes and
for reading from or writing to the MPU's data bus. If these facilities

are utilised we may transfer bytes directly between the MPU and external devices without the aid of a PIA or similar system.

The availability of address bus signals and the precise control bus signals varies somewhat between one micro and another. (For example the PET has partially decoded addresses available at its memory expansion port, as does the Apple at its I/O slots. These two 6502 machines produce a phase 2 clock signal to synchronise byte transfers on the data bus, while the Spectrum and other Z80 based micros generally have all 16 address lines available and MEMx and IOx lines to control timing.) Rather than cover all possibilities we will assume that no address decoding is performed within the computer (the simplifications introduced if partially decoded address are available should be apparent) and that the R/W and phase 2 clocking signal technique is used to synchronise operations.

First it is necessary to point out that if particular addresses are to be used to initiate data byte responses from an interface, the those addresses must not produce responses from elsewhere in the computer (ie. ROM, RAM or PIAs). In micros using a phase 2 clocking signal it is usual to find some addresses unused (eg. the spare ROM sockets of the PET, and the I/O slot address blocks in the Apple). In most cases a continuous group of addresses is available, perhaps 16, 256 or even 4096 consecutive addresses. For Z80 based micros holes in memory are not required (which is just as well, as there sometimes aren't any) if the RD, WR and IORQ control lines can be used for interfacing, as no clash with memory access will occur. However, some manufacturers use the IO control lines and the address bus to handle keyboard and peripheral functions (eg. the Spectrum), assigning individual address lines to particular roles without any decoding. This reduces the number of address lines available for external interfacing by the technique discussed below, and in practice the address lines A0 - A5 should be kept high if the Spectrum is to be interfaced in this manner. (For the Spectrum we have used address lines A8 - A11 for decoding, A0 - A7 = 127, and A12 - A15 = 0.)

We start by assuming that we have identified a block of 16 consecutive addresses which can be used to address an interface system, and that these addresses cover the range 0 - 15 on the least significant address lines, A0 - A3. Decoding A0 - A3 is, of course, quite straightforward using a 1 of 16 decoder. However, we also need to verify that this block of 16 addresses has been addressed by the MPU, and this requires us to check the values of the other 12 address lines, A4 - A15. This is most readily achieved using digital comparators, logic circuits which give a lo output if two multi-bit digital inputs are identical, and a hi output otherwise.

The 74C933 digital comparator operates on two 7 bit inputs and has two "comparison enable" inputs which must both be lo to enable comparison to occur. The circuit of fig 6.20 illustrates how two of

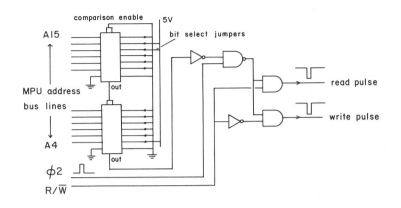

Fig 6.20 Partial decoding of the top twelve bits of the MPU address
         bus using 74C933 digital comparators. Note that either a read
         pulse or a write pulse is produced when an address match is
         present, depending on the level of the R/W control bus line.

these devices can be used to generate a pulse output while an address
(preset by the bank of switches or jumpers) is present on lines A4 –
A15 of the address bus. The pulse timing is derived from the phase 2
clocking signal of the control bus by ANDing the address valid signal
with the phase 2 clock. The resulting pulse is then ANDed with the R/W
level and its inverse to produce a "read" pulse or a "write" pulse –
depending on the R/W level. This enables us to use the 16 available
addresses for 16 input functions and 16 output functions
independently. (It is not a good idea to pulse the enable pin of the
digital comparator with the phase 2 clock. This is a relatively slow
CMOS device and the output pulse would be significantly delayed with
respect to the phase 2 clocking signal.)

    If the part address set by the bit select jumpers in fig 6.20
corresponded to 1001000.0000xxxx (where the xs are the least
significant 4 bits and are not used in this part of the circuit),
which is equivalent to a decimal number in the range 45056 – 45072,
then a read pulse would be generated by
                    X = PEEK(45056)   etc.
and a write pulse by
                    POKE 45056,X   etc.
(or the equivalent load and store instructions in machine code)

    These pulses may now be used to enable (one of) a pair of 1 of 16
decoders connected to address lines A0 – A3 as shown in fig 6.21. The
enabled decoder will produce on one of its 16 output lines a pulse
synchronised with the phase 2 clock (although slightly delayed as a
result of the combination of gate propagation delays). A normally hi

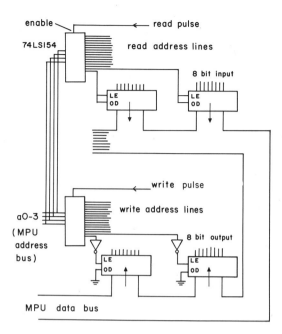

Fig 6.21 Decoding bits 0-3 of the MPU address bus in conjuction with
the read/write pulse produced in fig 6.20 allows control of
16 input and 16 output fuctions.

output decoder such as the 74LS154 (CMOS devices would be rather slow
at this point - the decoding time of the 4515B is about 800ns.)
produces a negative going pulse which can be used to operate the
output disable line of one of the TRI-STATE latches connected to the
MPU's data bus using the leading edge of the read pulse. This allows
the data to be loaded on to the data bus in good time for latching by
the computer at the end of the read pulse. In this manner
X=PEEK(45056) both selects the appropriate latch in the interface
circuit and transfers the data byte into the computer (storing it into
a BASIC variable, X) quite rapidly (a few microseconds if machine code
is used).

The trailing edge of a write pulse is used to latch data from the
data bus (in line with the memory timing discussed in chapter 5), so
the negative-going pulse from the decoder must be inverted as shown.
Thus at the end of the pulse the latch collects the byte loaded onto
the data bus by the computer's POKE(45056),X instruction.

It is likely that some of the addresses decoded would be used for
control functions in a complex interface system, just as they were in
the dual counter/timer system of fig 6.14. Others may be used to check
for busy signals by gating a read pulse with a busy signal output

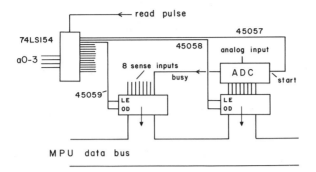

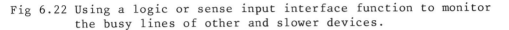

Fig 6.22 Using a logic or sense input interface function to monitor
         the busy lines of other and slower devices.

using a TRI-STATE latch as illustrated in fig 6.22. In this case the
state of an ADC can be tested until conversion is complete, and then
the data read. An example of some suitable coding is:

```
100 Y=PEEK(45057) :REM initiate conversion
105 REM check ADC busy signal
110 IF (PEEK(45059)AND1)<>0 THEN 110: REM check again
120 X=PEEK(45058) :REM read ADC into X
200 REM DATA BYTE NOW IN X
```

Note that the convert signal was sent by addressing 45057 using a
PEEK, although no useful data would be stored in Y. We could equally
have used a spare write line and POKE 45057,0 to initiate conversion.

6.7 Dynamic interfaces

     Many of the signals measured or produced by laboratory
instrumentation cover a wide range of values. In manually controlled
instruments this often necessitates the selection of an input or
output range before the start of an experiment. For example, in
recording a chromatogram it is usual for the operator to select the
sensitivity range of the detector before injecting the sample, and,
unless the sample is of a routine nature, the initial choice of range
is a matter of guesswork. If the guess is wrong and the peaks are
either too small or off-scale, one can always change the range and
inject another sample. When a computer is used to monitor a signal it
becomes possible to program the computer to select the optimum signal
conversion characteristics and to adjust these (if necessary) as the
experiment proceeds. A signal converting interface which can have its
conversion characteristics changed under computer control can be
called a dynamic interface.

     To illustrate the value of a dynamic interface we will consider

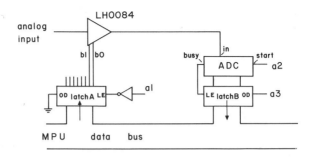

Fig 6.23 A analog input interface with a conversion gain digitally
         selected as 1, 2, 5 or 10 from the user's program.

the problem of monitoring an analog signal in the range 0-10 V, where
the signal value is required to a precision of 1% whatever its value.
Clearly a simple 8 bit analog input interface would not be appropriat
for such a task as it would be capable of a resolution equivalent to
only 50 mV (allowing for +/- 1 LSB conversion errors) no matter what
the value of the signal. So a 100 mV signal would be converted with
only a 50% precision. There are numerous solutions to this problem
although for brevity we will consider only two of the possible
techniques.

    Figure 6.23 shows a dynamic analog input interface for the
conversion of a single analog signal in the range 0-10 V into an 8 bi
digital signal. Addressing the different requirements of the interfac
can be performed using any of the techniques discussed in this chapte
and no specific addressing circuits are included in the figure,
although the version constructed in the author's laboratory used the
synchronous byte transfer technique described in section 6.6. The
circuit consists of an amplifier stage formed by a digitally
programmable instrumentation amplifier (LH0084) which provides a gair
of 1, 2, 5 or 10, depending on the digital value of a 2 bit gain
control code (ie. 00, 01, 10 or 11 respectively). This stage is
followed by an 8 bit ADC, and for simplicity we will assume that this
device produces 255 for a 10 V input. The latch A is latched by
addressing the interface with address a1 (eg. 45057), the ADC
conversion cycle is started by address a2 (45058), and the ADC output
(latched into latch B by the ADC's busy signal) is read by addressing
with a3 (45059).

    The 8 bit digital value,X, read in by the computer corresponds
to

$$X = 255*G*V_1/10$$

where V_1 is the analog input voltage whose value is required and G is

the gain of the amplifier. When the analog input is, say, 8 volts and
G=1, X will be about 204 and the precision (+/- 1 LSB) corresponds to
0.5%. When V_1=0.8 V and G=1 X would be about 20 and the precision
about 5%, but G can be increased to 10, so that X becomes about 204
and the precision 0.5% again. To maintain a reasonable level of
precision we require only that the byte input routine monitors the
value being read and modifies the gain (when possible) if the byte
value falls below 100 or rises above 200. A BASIC version of a simple
routine for operating the interface is given below. The routine relies
on the slow speed of BASIC to allow the ADC conversion to complete
before latch B is read. To prevent oscillatory behaviour when readings
are close to 100 or 200, the gain change tests (lines 30 and 40) use
test values of 90 and 220 rather than 100 and 200. Note that the
structure GOSUB 200: GOTO 20 is included for clarity; faster operation
is achieved by using GOTO 200 and changing line 260 to GOTO 20.

```
10 REM SUBROUTINE TO READ ANALOG INPUT
11 REM G(AIN) INITIALLY 1 AND GC (GAIN CODE) 0
20 GOSUB 100 : REM READ ADC INTO X
30 IF X<90 AND G<10 THEN GOSUB 200:GOTO20
40 IF X>220 AND G>1 THEN GOSUB 300:GOTO20
40 V1 = X*10/(G*255) : REM V1 = ANALOG VOLTAGE
50 RETURN
100 REM READ ADC and time
110 POKE 45058,0: REM START ADC
120 GOSUB 400 : REM reads clock (not listed)
130 X = PEEK(45059): RETURN
200 REM INCREASE GAIN: GAIN CODE IS GC, INITIALLY 0
201 REM      0=<GC=<3
210 GC = GC+1
220 IF GC>3 THEN GC = 3
230 IF GC<0 THEN GC = 0
240 POKE 45057,GC: REM SET CONVERSION GAIN
250 G = GA(GC): REM GA(I) HOLDS 1,2,5,10 FOR I=0 to 3
260 RETURN
300 REM DECREASE GAIN
310 GC = GC-1:GOTO 220
```

An improved technique which could be extended to cover a much
wider dynamic range is illustrated in fig 6.24. This circuit utilises
three 725 instrumentation op-amps which do not have the precision of
true instrumentation amplifiers. Of course a similar approach could
also be adopted with selectable gain instrumentation amplifiers. In
fig 6.24 the analog input signal is offset by a dc voltage generated
by an 8 bit DAC (operated with b0-b2 grounded, so the offset is
applied as one of 32 values in the range 0 and 10 V), and then
amplified by an inverting configuration *10 amplifier. The signal is
the amplified again by a second inverting configuration amplifier,
with a gain determined by one of three feedback resistors connected by
an analog switch IC and trimmed to be *3.3, 10 or 33. The DAC output

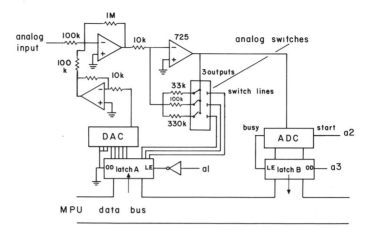

Fig 6.24 A wide range dynamic analog input interface selectable offset
and conversion gain. Operates over the range 0-10 V with a
precision of better than 1 mV.

is determined by b3-b7 of latch A, while the amplifier gain is
selected by b0-b2 of latch A. The remaining functions are similar to
those of fig 6.23.

The byte value, X, read by the computer now becomes

$$X = 255*(V_1-OS)*G/10$$

where V_1 is the analog input signal,
      OS is the dc offset derived from the DAC, and
      G is the overall gain of the amplification stages.

The arrangement illustrated allows both the offset and the gain to be
set using a single byte latched into latch A. The byte value can be
determined by adding a gain code (1, 2 or 4) to 8 times the required
offset (0-31). The dc offset voltage would be (0-31) times 10/31 volt
for a 10 V output DAC, although this value could be modified by
altering the inverting amplifier's circuitry. As shown the circuit
allows the offset to be set to within 300 mV of the analog input, and
the difference signal to be amplified up to 330-fold. Thus the
precision of the calculated input value should be better than 1 mV,
assuming that appropriate care is taken in trimming the op-amp offset
and gains. In practice we found that errors of up to 3 mV could
occur.

Not surprisingly it is possible to obtain commercial dynamic
analog input interfaces with considerably better performance. If a
very wide dynamic range and high accuracy are important, then one of

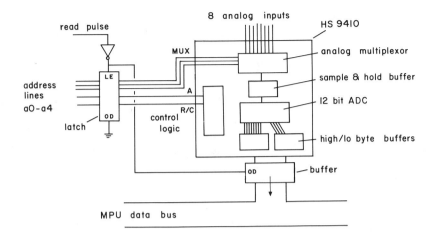

8 analog inputs

read pulse

HS 9410

MUX

analog multiplexor

address
lines
a0-a4

LE

A

sample & hold buffer

12 bit ADC

OD

R/C

latch

control
logic

high/lo byte buffers

OD

buffer

MPU data bus

Fig 6.25 A multiplexed input analog input interface allowing the
reading of up to 8 analog signals and 12 bit ADC.

the autoranging multidigit digital voltmeters fitted with a standard
interface system for connection to a computer offers a versatile and
reliable solution.

While the dynamic interfacing described above was illustrated
using an 8 bit ADC, in practice a higher resolution ADC would be more
appropriate. High quality high resolution ADC are moderately expensive
and in systems where several analog inputs are to be monitored it is
generally desirable to switch the analog input to the ADC from several
sources, rather than have an ensemble of separate ADCs monitoring all
the inputs. Switching analog inputs can be carried out using analog
multiplexor ICs, but there are a number of ADC circuits on the market
with built-in multiplexors. Several of the multichannel ADCs are
intended for direct connection with microprocessors and are equipped
with MPU compatible address, data and control lines (some are able to
provide signals to generate MPU interrupts at the end of conversion).
We will consider one such device, although discussing its connection
to a microcomputer using the types of addressing described earlier in
this chapter.

The part circuit of fig 6.25 shows the HS9410 (Hybrid Systems
Corporation) 8 channel, 12 bit ADC connected as an analog input
interface using the synchronised addressing method. Any of the other
methods would be equally suitable for the limited use we make of this
device's facilities, as the (12 bit) conversion time is about 30
microseconds. The device requires 5 inputs: 3 address lines to carry a
code for selecting 1 of the 8 analog inputs; a "read/convert" line
(R/C), which enables the digital output when hi, disconnects the
digital output when lo (ie. the output go to the high impedance

state), and initiates the ADC conversion on a hi-to-lo transition; an
a "device address" line (A) which selects 12 bit conversion if lo and
8 bit conversion if hi when R/C receives a hi-to-lo transition, and
selects which byte is output on the data lines when R/C is hi, (A lo
gives the most significant byte, A hi gives the least significant byt
- including 4 trailing zeros).

In this implementation the circuit is addressed by read cycle
instructions, ie. BASIC X=PEEK(addr.) or assembler LDA instructions,
using 5 lines of the MPU address bus. Thus the read pulse derives fro
a circuit which decodes the MPU address lines a5-a15, one less than
illustrated in fig 6.20. If the required 32 consecutive addresses
start at B, then the following BASIC commands may be used to control
the operation of the circuit:

```
10 X=PEEK(B+1+4*N)     select analog channel no. 0=<N>=7
20 X=PEEK(B+4*N)       initiate 12 bit conversion cycle
30 H=PEEK(B+1)         read high byte of result
40 L=PEEK(B+3)         read low 4 bits of result
50 V(N)=(16*H + L/16)*10/4096 volts on channel N
```

An 8 bit conversion may be carried out using:

```
10 X=PEEK(B+3+4*N)     select analog channel no N
20 X=PEEK(B+2+4*N)     initiate 8 bit conversion
30 V(N)=PEEK(B+1)*10/255    read byte value
```

If the equivalent assembler language instructions are used to
implement these operations then a delay is required between 12 bit
inititiation and the high byte read (about 30 microseconds), and
between the 8 bit inititiation and the single byte read (about 21
microseconds). The delay may be programmed with a suitable NOP loop.
The manufacturer's data sheets shows how conversions may be
"overlapped" to achieve a very high data aquisition rate.

CHAPTER 7

STANDARD INTERFACE SYSTEMS

Interface standards exist so that units from one manufacturer can be connected to devices from another manufacturer and still allow digital signals to be passed from one to the other. In the computer context a standard interface system can be viewed as consisting of the elements illustrated in fig 7.1. The "system" actually consists of two interface units - the computer interface and the device interface. To allow for digital signal transfer between the two, the "standards" are primarily concerned with the characteristics of the signals involved - their number, signal levels, impedances, timings etc. - rather than with the actual electronic circuits in either interface, although some characteristics of the latter may be important.

7.1 Introduction

A growing number of standard interface systems is available, but it is important to understand that some of the most widely used standards were not actually devised for the convenience of the microcomputer users (or manufacturers). Some were developed for use in technical instrumentation, to allow one piece of electronics to control another, etc. Others are primarily communications standards, developed for the transmission of non-technical messages over intermediate distances (eg. using teleprinters). Those that have been adopted in the microcomputer world are not necessarily ideally suited to the task, but they have dramatically increased the number of devices that can be "plugged in" to suitable microcomputers. They have also reinforced the value of the standards to the extent that most

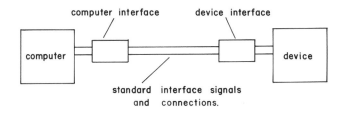

Fig 7.1 The elements of a standard interface system.

microcomputer peripherals and many laboratory instruments now
incorporate a device interface conforming to one of the major
standards.

    In this chapter we will examine two of the most popular standard
interface systems, one (the IEEE 488 standard) intended for use with
laboratory instrumentation, and one (RS232) primarily designed as a
communication standard - but unquestionably the most widely used micro
input/output standard. Many microcomputers are fitted with either one
or both of these standard computer interfaces, or may have such
interfaces added to their hardware. The advantages of standard
interface systems are that

a) if one buys a new computer it is not necessary to buy a whole new
   range of interfaces for it - only one standard computer interface
   is required,
b) a large range of instrumentation with standard device interfaces
   built in is available from dozens of different manufacturers, and
c) in some cases the program control of the standard interface system
   can be exercised by a couple of simple BASIC statements (ie. no
   PEEKing or POKEing is required).

Unfortunately there are a couple of disadvantages which have to be
weighed against the above, one of which is the high cost of the
standard device interface on simple devices (such as analog input or
pulse counter interfaces) compared with the simple parallel port
system described in chapter 6. The second is that the implementation
of at least the best known laboratory instrumentation standard has
become bifurcated to the extent that the units from one manufacturer
will often just not work with computers from another.

7.2 The IEEE 488 standard

    The best known standard interface system for use with laboratory
instrumentation is the IEEE 488 standard, developed originally by
Hewlett Packard and often known as the Hewlett Packard Information Bus
(HPIB) system by Hewlett Packard or the General Purpose Information
Bus (GPIB) by everybody else. (In its present form the standard is
IEEE Std 488-1978. The full specifications are available from: IEEE
SERVICE CENTER, 445 Hoes Lane, Piscataway, New Jersey 08854, USA.,
and helpful documentation is available from Hewlett Packard. This
standard is identical with ANSI MC1.1.). The standard covers the form
of data transmission (8 bit parallel, byte serial), the maximum byte
transmission rate (1 Megabyte per second - although no microcomputers
can actually manage that), the signal levels (TTL compatible), the
length and number of wires used for data lines and control lines, even
the stackable connectors used on the 24 way cable between instruments
(A closely related standard is the IEC 625-1, which calls for a
different style of plug and 25 way cables between units. Adaptors are
available for interconnecting the two systems, from N.V. Philips and

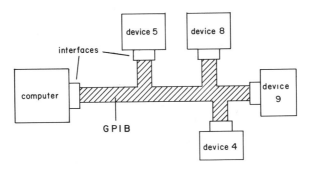

Fig 7.2   The connection of several devices to a GPIB. Devices are
          addressed using a device address and only the addressed
          devices talk or listen.

A.G. Siemens among others.). As a small aside it is interesting to
note one of the problems encountered by international standards. The
stackable connector of the GPIB uses pillar bolts to secure the
mechanical connection. Some of these standard connectors have black
bolts, and these have metric threads. Other connectors have silver-
coloured bolts, and these have imperial threads. If these two
different threads are inadvertently screwed together they tend to
become jammed and to damage each other's threads.

     A full description of the GPIB system is beyond the scope of this
text, but a survey of those aspects most apparent to the user is
included because some computers (eg. the PET/CBM range and the Sirius)
have I/O ports which are compatible with the IEEE 488 standard and
software to handle data transfers, while others (eg. the Apple and HP
ranges) can have IEEE 488 systems added at relatively low cost. A
number of manufacturers now produce GPIB compatible interfaces of the
types described earlier, and a range of instrumentation fitted with
GPIB interfaces, from digital multimeters and signal generators to
storage oscilloscopes and graph plotters, is available from a large
array of manufacturers.

     The most important feature of the GPIB system is that up to 15
device interfaces may be connected to a common bus, as illustrated for
a typical system in fig 7.2, although where a large number of devices
is involved it is better if the devices are connected by "daisy
chaining" in order to minimise the capacitance of the bus wiring. Most
GPIB inter-unit cables are 1 or 2 metres in length, and the standard
specifies a maximum length of 4m (although 8m cables are available!)
and a maximum total length of cabling of 20m.

     Communication over the bus is controlled by one of the devices
connected to the bus and known as the Controller. Although other

arrangements are possible, when one of the popular microcomputers is
used to communicate with a number of devices via the GPIB, the
microcomputer is used as the controller of the bus, and generally is
the only controller which may be connected to the bus - thus it is not
usually a simple matter to connect a number of micros together using
the GPIB. (These limitations arise because most microcomputer
implementations of the GPIB apart from Hewlett Packard's allow only
for a subset of the standard communication signals.) When the computer
needs to communicate with a specific device attached to the bus a
device address is used, each interfaced device being referred to by a
device address which can be selected by means of switches on the
interface (sometimes not easily located or altered in low cost units).
Only the addressed device interface responds to data or requests for
data from the controlling computer.

Of fundamental importance in understanding communication over the
GPIB is that all signals tansmitted on the bus are based on negative-
true logic, including those signals which represent data bytes. Thus a
lo level is to interpreted as a 1, and a hi level as a 0. This
relatively unusual arrangement will soon become apparent as we discuss
the principal features of the bus. All lines idle at the hi level and
all devices connected to the bus must allow other devices to assert a
low level on any line. Thus while bus electrical signals are all TTL-
type, connections to the bus may only be made via open collector or
TRI-STATE outputs.

Interfaces which can only place data bytes on to the bus (eg. a
simple ADC) are called Talkers in GPIB parlance, whereas those which
can only accept data bytes from the bus (eg. a simple DAC) are called
Listeners. More sophisticated devices which have the ability to act as
both Talkers and Listeners are sometimes called Intelligent (although
their intelligence may be somewhat limited). A typical example is the
Hewlett Packard model 3455A multimeter, which may have its operating
fuction (eg. voltage, current, resistance measurement, etc.) and range
(eg. 100 mV, 1 V, 10 V etc.) set with bytes from the computer, and can
output readings to the computer in the form of bytes representing
ASCII coded characters (eg. 55.20, 1.2345 V, etc.). Of course a
controlling computer can act as a Talker or Listener in addition to
being the Controller of the bus.

7.2.1 The bus conductors

The interdevice connection of the GPIB is illustrated in fig 7.3.
The bus actually consists of three sub-busses: a data bus of eight
conductors (dio1 - dio8) plus one ground, a three conductor control
bus which deals with the handshaking of bytes along the data bus (plus
three associated ground conductors), and a five conductor management
bus used for passing control signals between devices on the bus, and
for differentiating between the function of bytes being passed on the

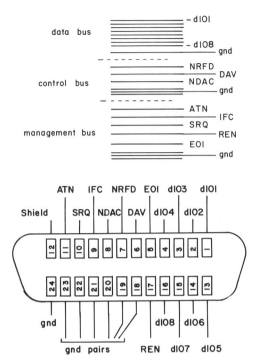

Fig 7.3   a) The sub-busses of the GPIB and the nmemonics used to
identify each conductor. b) The IEEE standard connector
ribbon assignments (device socket view).

data bus (only three of the management bus lines are provided with
their own ground conductors). All bus signal levels are TTL
compatible.

The control bus lines are known as "Not Ready For Data" (NRFD),
"DAta Valid" (DAV) and "Not Data ACcepted" (NDAC). These names, some
of which tend to trip rather than roll off the tongue, help to remind
us that the GPIB system is one of the important instances of negative
logic or lo=true logic. Thus when the NRFD conductor is held lo by
anything on the bus, the bus is not ready for the transmission of
data. Only when the NRFD line is hi can data transmission start. The
control bus conductors must be connected (and must be operational)
before any byte transfer can take place over the GPIB. In section
7.2.2 we shall discuss the role of the signals handled by the control
bus, and in section 7.2.6 we shall consider an economical way of using
the control bus to handshake bytes to and from non-standard
interfaces.

The five management bus lines are also known by abbreviations:

ATN (attention!) The Controller sets this line lo to indicate to all
    devices on the GPIB that the data bus is carrying an address or a
    control byte (as distinct from a data byte).
IFC (interface clear) The Controller sets this line lo to force
    devices attached to the bus to leave the bus in a standardised
    state (usually with all attached devices idle and waiting for a
    signal from the Controller).
SRQ (service request) Any GPIB device fitted with the service request
    function can set this line lo to indicate to the Controller that
    it needs servicing - eg. it wants the Controller to communicate
    with it.
REN (remote enable) Used by the Controller to switch instruments
    fitted with this function from "local" (ie. front panel) control
    to "remote" (ie. GPIB) control.
EOI (end or indentify) The "end" function is used (optionally) by
    Talker devices to indicate that a data byte being transferred is
    the last data byte of a sequence, the EOI line being held lo
    during this byte transfer. The "identify" function is somewhat
    rarer, and is used by the Controller during a parallel poll (a
    sort of roll call in which up to eight devices can signal their
    presence by holding individual data lines lo) of devices present
    on the bus.

It is possible to use a GPIB connector on a micro to communicate
with a device interface without any connection to the management bus.
However, if several devices are connected to the bus it is usually
desirable to operate them by addressing the individual units, which in
turn requires that the ATN line is used. ATN and the three control bus
lines (NRFD, NDAC and DAV) form a set which can be regarded as the
minimum configuration for controlling the transfer of data over the
data bus. Some micros equipped with GPIB interfaces do not monitor or
use all of the management bus lines, and several commercial GPIB
device interfaces intended for use with microcomputer peripherals
implement only ATN and EOI.

7.2.2 The byte handshaking sequence

The byte handshaking sequence illustrated in fig 7.4 occurs for
every byte transferred, whether it be a data byte or an address or
command byte. The sequence begins with the source device (which may be
a Talker or Controller) allowing the DAV line to go hi (remember the
GPIB uses negative logic, so hi is false and in this case indicates
that data on the data bus is not valid) and all acceptor devices on
the bus holding the NDAC line lo - indicating that they have not
accepted the current content of the data lines. The source device then
places the byte on the data bus and waits for all the devices on the
bus to indicate that they are ready for data by allowing the NRFD line
to go hi. When the NRFD line is hi the source forces the DAV line lo,

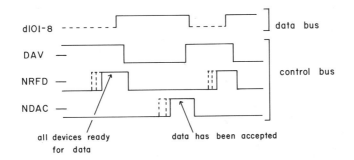

Fig. 7.4 Level changes on the control bus during the handshaking
          sequence of the GPIB. The vertical broken lines indicate the
          different response times of devices on the bus.

to indicate that valid data is on the data bus and ready for
collection, and all acceptor devices on the bus repsond by holding
NRFD lo - indicating that they are no longer ready for new data. There
is then a time interval during which the acceptor devices have the
opportunity to read the data lines. All devices should respond by
releasing the NDAC line, so that it goes hi when the last device (ie.
the slowest device) releases it. When this happens the source releases
DAV so that it returns hi, indicating that the data lines no longer
hold valid data, and the acceptor devices respond by holding the NDAC
line lo. At this point one byte has been transferred from the source
to all acceptor devices on the bus. In fig 7.4 the broken lines
illustrate different response times which occur when several acceptors
are connected to the bus.

     A couple of points need to be noted in connection with the
handshaking sequence. Firstly, all devices connected to the GPIB are
required to permit the handshaking of the bus whether they have been
addressed or not (see below). This means that the devices must either
participate in the handshaking, or must allow the lines to have their
levels determined by other devices. The IEEE 488 standard does not
specify the types of circuits used to send or receive signals over the
bus conductors, and in practice some GPIB devices use TRI-STATE[R]
circuits while others use open collector circuits whose outputs are
connected via resistors to ground and a +5 V supply. The two types are
distinguished by the GPIB implementation codes E1, for open
collectors, and E2 for TRI-STATE[R] buffers - which are capable of
faster operation, typically 0.5 Megabytes per second. A consequence
of this is that some GPIB devices may not allow bus lines to return to
5 V unless the power to that device is switched on, so that having one
or more unpowered devices connected to a bus can prevent the bus being
used. This can result in a certain amount of confusion as other
devices can be left connected to the bus while unpowered and allow
data transfer between other device to continue undisturbed.

Secondly, devices differ in the amount of time they allow other
devices to respond to signals on the ATN or the control bus lines. On
the PET computers, for example, 14 microseconds are allowed for
devices to respond (by setting NRFD and NDAC lines hi and lo
respectively) to a lo on the ATN line, while the IEEE 488 standard
specifies only 200 ns. Similarly the PET allows 64 ms for a device to
respond to DAV lo with a NDAC hi, whereas other Controllers allow
different times – some shorter, others being prepared to wait
indefinitely. The timing requirements of many possible bus activites
are too complex for detailed discussion in the present context, but
the user should be aware that many microcomputers are actually slow
devices by GPIB standards so that their use with some fast GPIB
devices can give rise to difficulty. The best advice is test before
you buy, as manufacturers tend not to be interested if one of their
products will not communicate with a device from another manufacturer.
Note that there is no restriction on the time allowed for NRFD to go
hi (except in response to ATN lo), and the NRFD line is therefore used
to hold up data transfers when an acceptor is busy.

7.2.3 Controller signals

The bus Controller, which is the microcomputer in the present
context, always initiates the transfer of data bytes over the bus.
This is true whether the computer is to act as the Talker (ie. sending
the data bytes) or the Listener (ie. receiving them), or even when the
computer takes no further interest in the transfer – such as when the
bytes are to be sent from an instrument directly to a printer or a
disk drive. The Controller initiates data transfer by taking the ATN
line lo and, when all other devices on the bus have responded by
letting NRFD go hi and NDAC go lo, placing a byte on the data lines.
The Controller acts as a source during this operation, and when ATN is
held low all other devices on the bus act as acceptors. The byte (or
rather bits 0-6, as bit 7 is not used while ATN is low) placed on the
bus using the convention that hi=0, lo=1, can be one of six types of
"command" to the interfaces/devices connected to the bus:

A bus command. Bits 5 and 6 of the byte are 0s, and bits 0-4 carry a
     code (0 - 31) which the devices on the bus understand, eg. a GET
     (Group Execute Trigger) command – which should not be confused
     with a BASIC GET, see section 7.2.5.
MLA (My Listen Address, also called LAG, Listen Address Group) Bit
     6=0, bit 5=1. Bits 0-4 carry an address (0-30, not 31) which
     activates the device(s) with the appropriate address so that it
     becomes an active Listener, ready to accept data from the bus.
     Note that all devices handshake data bytes, but only the addressed
     Listener(s) actually read them.
UNL (Unlisten) The byte X0111111 causes all active Listeners to be
     deactivated.
MTA (My Talk Address, also called TAG, Talk Address Group) Bit 6=1,
     bit 5=0. Bits 0-4 carry an address (0-30, not 31) which activates

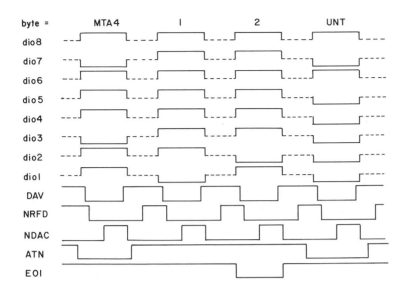

Fig 7.5   A data transfer initiated by the Controller and carried out
by the addressed talker. At the end of the transfer the
Controller issues an Untalk instruction which releases the
talker.

the device(s) with the appropriate address to become an active
Talker, ready to transmit data over the bus. The activation of a
Talker automatically causes the deactivation of any previously
active Talker.

UNT (Untalk) The byte X1011111 causes the active Talker to be
deactivated.

MSA (My Secondary Address, also called SCG, Secondary Address Group)
Bits 5 and 6 are 1s and bits 0-4 carry a secondary address (0-31),
a code which may be used for the transfer of a command to the
addressed device (such as the mode in which it should operate, or,
in the case of a printer, the typeface or colour with which it
should print, etc.). Relatively few of the low cost GPIB devices
make use of secondary address facility, although Commodore
peripherals for PET computers use them extensively.

When the Controller wishes to initiate a data transfer sequence
it outputs a MLA or MTA code (depending on whether the device
addressed is required to act as a Talker or a Listener) on the data
bus while holding the ATN line of the Management bus lo. For example,
initiating data output from a Talker device with address 4 requires
the Controller to output the byte 01000100 with ATN lo as shown on the
left hand side of fig 7.5. The Controller may the either return ATN hi
so that data transfer can begin, or may retain ATN lo while it outputs

further bytes containing perhaps an MSA code or an MLA code (if the data is to be recieved by some other Listener device), before returning ATN hi. When ATN returns hi the addressed Talker device begins to handshake data bytes onto the data bus, again using the control bus to control the handshaking as illustrated in fig. 7.5, where the Talker transmits byte values of 1 and 2. The broken lines indicate that the level on the data lines is irrelevant, the actual states will depend on the type of bus drivers used.

During the handshaking of the last data byte the Talker may set EOI lo, although not all GPIB devices do this - so if a micro expects EOI to be lo to terminate the transfer of a multi-byte string of data then this could lead to trouble, (I am not aware of any micros which actually require this at present). At the end of the data transfer sequence the bus remains idle until the Controller initiates some new action. In most cases the controlling microcomputer has been either the Listener or the Talker during the data transfer, and it is usual for the Controller to deactivate the other device(s) as soon as the transfer is over. The Controller does this by issuing an unlisten (UNL) or untalk (UNT) command (depending on the nature of the completed data transfer) by holding ATN lo and handshaking 00111111 or 01011111 respectively over the bus. However, not all microcomputer (or minicomputer) Controllers do this at the end of every data transfer (the IEEE standard does not lay down rules of behaviour for data transfers) and this is one of the areas of difficulty when connecting devices from different manufacturers.

7.2.4 Typical BASIC data transfers

While the handling of data, control and management busses can be performed at the assembler level, either by writing the appropriate code or by using routines of the computer's operating system (if available), it is generally most convenient to communicate with GPIB devices using a BASIC program. Many of the commercial GPIB devices transmit and receive data in the form of ASCII character strings, so BASIC is convenient for handling communications. Relatively simple BASIC-type commands have been built in to GPIB-equipped micros and these can be used to transfer data bytes between the computer and devices on the bus, and the operation of the management and control busses are completely transparent. However, some understanding of the system is still desirable because of the specific requirements of many GPIB devices, and becomes essential if required command functions (ie. bytes transmitted with ATN lo) are not implemented in BASIC.

One of the principal areas of incompatibility between the different dialects of BASIC available on popular micros concerns the input/output systems. Thus PET BASIC uses INPUT#, others use IN and others READ, to input data bytes through GPIB connectors. In discussing the sequence of events associated with BASIC input/output using the GPIB we shall confine ourselves to PET BASIC. This should

Table 7.1   Examples of byte transfers on the GPIB
             using PET BASIC

```
10 OPEN 5,9          :REM opens channel 5 to device 9
20 GET#5, X$         :REM input a single byte from channel 5
30 INPUT#5, X$       :REM input an unspecified number
                          of bytes until a <cr> is input
40 INPUT#5, X        :REM input ASCII version of a decimal
                          number, translate into the numerical
                          value & store in X

60 OPEN 7,6          :REM open channel 7 to device 6
70 PRINT#7, "0"      :REM output a single byte via channel 7
                          (ASCII 0 = decimal 48)
80 PRINT#7, X$       :REM output the bytes in X$ + <lf>+<cr>
90 PRINT#7, X        :REM output the ASCII codes for the
                          decimal value of X
```

---

not prove restricting as the translation to other dialects is
relatively simple and the GPIB signal sequences are broadly the same
for all common micros (although differing in detail).

There are three types of byte transfer instructions in PET BASIC,
two for the input of data (INPUT# and GET#) and one for output
(PRINT#). These instructions are all used with associated channel
numbers, the channel numbers themselves being defined in OPEN
instructions where they may be identified with particular device
addresses and (optionally) secondary addresses. (Note: BASIC4 also
provides a DOPEN instruction for use with the GPIB.) An example of
data input from a Talker device using BASIC on a PET Controller is
given in table 7.1, where it should be noted that the GPIB device
address appears only in the OPEN channel instruction – subsequent
instructions referring to channel numbers which are used in the form
of an index by the PET. In this example the GET#5,X$ instruction
results in the input of a single byte from the Talker followed by its
storage in variable X$. On the other hand the INPUT#5,X$ instruction
causes the input of an unspecified number of bytes, the transfer being
terminated by the transfer of a byte with the decimal value 13 (the
ASCII code for "carriage return", used on micros for "return" or
"enter", see appendix 2). On the PET micros INPUT# can input up to 80
bytes before finding a carriage returr, but more than 80 bytes will
cause an error (not a standard GPIB characteristic).

The bus signals involved in these two transactions are shown
diagramatically in fig 7.6. In the case of the GET#5,X$ instruction
(fig 7.6a) the PET Controller actually outputs an MTA=9 command (the
device address 9 having been specified in the OPEN 5,9 instruction of

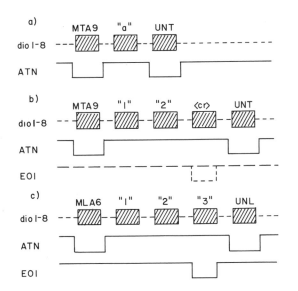

Fig 7.6   The GPIB activity typical of a) GET#, b) INPUT# and c) PRINT#
          instructions on the PET microcomputer. The control bus
          handshaking signals are not shown, and the EOI signal in b)
          is dependent on the device interface.

table 7.1), configures itself as a Listener and receives a single byte
from the Talker, then becomes a Controller again and outputs an UNT
command. In our example the byte transmitted is the ASCII code for
"a". (Unfortunately PET ASCII codes are not the same as true ASCII
codes, the upper and lower case alphabetic characters being confused
for example. Thus if a GPIB Listener device actually requires lower
case alphabetic characters some code conversion is needed, or CHR$
functions may be used.) In the case of the INPUT#5 instruction (fig
7.6b) the PET outputs MLA=9, configures itself as a Listener and
accepts a series of bytes until either one has the value 13 or the EOI
line is sensed lo (in which case a 13 is forced into the PET's input
buffer anyway and the PET's status byte, ST, is set equal to 6). At
this point the PET reverts to its controlling roll and outputs a UNT
command.

     The illustration in fig 7.6c shows the bus signals produced by
the BASIC instruction PRINT#7,"123". Again the PET starts as the
Controller, outputs MLA=6 (the device address 6 coming from the OPEN
7,6 instruction - see table 7.1), then acts as a Talker and outputs
bytes representing the PET ASCII characters 1, 2 and 3, with EOI being
held lo during the output of the 3. Finally it reverts to being a
Controller and outputs UNL, leaving the bus clear for its next
communication.

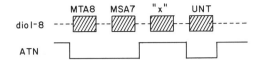

Fig 7.7   The bus activity typical of a PET BASIC GET# instruction in
          which the channel number refers to an OPEN instruction
          containing a secondary address (of 7). x represents the byte
          "got".

    In those cases where secondary addresses are used, the PET
input/output instructions transmit the secondary address defined in
the relevant OPEN statement immediately after the transmission of the
device (primary) address. A typical example is shown in fig 7.7, where
the bus signals illustrated are those associated with the BASIC GET#
instruction of table 7.1 but where the OPEN instruction is OPEN 7,8,7.

7.2.5 Bus commands

    While the majority of data transfers can take place under the
control of a BASIC program, it is important to realise that not all
micros have bus commands implemented via BASIC. Thus on several micros
bytes with bits 5 and 6 as 0s cannot by output directly from BASIC
while the ATN line is held lo. This is unfortunate as a number of
useful laboratory functions require just this facility. Two examples
of commonly used bus commands are:

Group Execute Trigger (GET, code 8) causes a group of devices (which
        were previously addressed to become active Listeners and which
        are capable of responding to GET commands) to take some
        simultaneous action - such as start taking measurements.
Device Clear (DCL, code 20) causes all devices capable of responding
        to go into a predefined state, typically with empty buffers
        ready for a new set of measurements.

Usually it is possible to write small assembler routines which will
effectively provide such functions via a simple BASIC SYS call. Some
companies even provide routines in plug-in ROM for different micros
(for example, Audiogenic produce an IEEE ROM for the PET which enables
all standard bus commands to be implemented with simple BASIC calls).

    Of course it is also important to appreciate that devices
labelled as GPIB compatible will probably incorporate only a limited
number of GPIB functions. With the revision of the IEEE 488 standard
in 1978 came the recommendation that all GPIB devices should be marked
near the GPIB connector with a set of codes to indicate which GPIB
interface functions the unit supported. A list of the available
nmemonic codes is given in table 7.2, although in use these are all

Table 7.2   The nmemonic codes used to indicate the
            interface fuctions supported by a GPIB device

nmemonic	function name
T,TE	Talker or Extended Talker
L,LE	Listener or Extended Listener
SH	Source handshake
AH	Acceptor handshake
RL	Remote/Local
SR	Service Request
PP	Parallel Pool
DC	Device Clear
DT	Device Trigger
C	Controller
E	Driver type.

followed by a numeric codes which indicate something about the
implementation. (A 0 means not implemented, 1 means full
implementation, and other values mean partial implementation.
Naturally there are exceptions - E1 means open collector bus drivers,
while E2 means TRI-STATE drivers). Extended Talkers and Extended
Listeners are devices which can use two bytes for addressing as
Talkers or Listeners, the second byte being transferred as a secondary
address code.

7.2.6 A non-standard system for the GPIB

    The advantages of being able to communicate with a variety of
devices on the GPIB using BASIC are so great that the availability of
a GPIB connection can govern the choice of microcomputer for a
laboratory. There can be little doubt that this factor was a major
contribution to the success of the PET computers in meeting small
scale laboratory microcomputer requirements. While it is not difficult
to construct GPIB interfaces of the kind described in chapter 6,
providing a specification which actually meets the IEEE standard can
be a time-consuming business. Many commercial interface systems of
high quality are avaiable and many of these utilise a microporcessor
to handle the GPIB communications. However, for many micro
applications going to such lengths is unnecessary, because the micro
itself is probably relatively slow and implements only a small subset
of the full IEEE standard in BASIC. Furthermore, even if an in house
interface unit is to be used on a bus which also has other high
quality units connected, most applications call for data transfer only
between one device and the computer at a time - so that, provided the
in house unit does not interfere with handshaking sequence, it is

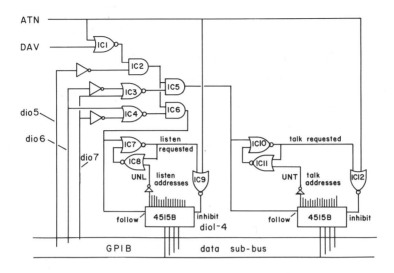

Fig 7.8   The address decoding portion of a multifunction interface
           which may be addressed using a GPIB. The handshaking sequence
           is carried out by the circuit of fig 7.9.

unlikely that the mixing of home-made and commercial units will lead
to problems.

     Units of the kind described below have been in use for some years
in the author's laboratory, connected to PET computers and various
commercial GPIB instruments, or allowing Commodore 64s to communicate
with GPIB devices. The GPIB laboratory interface unit described below
implements only a limited number of GPIB facilities but nevertheless
provide a simple, low-cost means of data collection and experiment
control using the types of interfaces described in chapter 6, operated
with GET# and PRINT# commands from PET BASIC. The GPIB adaptor
described in the following section provides a low-cost means of
allowing a non-GPIB micro to communicate with GPIB devices, although
the handshaking operations are performed under software control and so
require machine code routines to achieve the necessary speed.

     The circuit illustrated in fig 7.8 shows part of a multi-function
interface unit in which the various functions may be activated by
addressing the interface with addresses between 16 and 30, lower
addresses being available for use with other GPIB periperals. The
addresses are decoded from bits 0-4 of the MTA command using a 4515B
decoder for bits 0-3 (as described in chapter 6), activated only when
bit 4, dio5, is a 1. (Note the inverters in dio5-7. The GPIB data bus
lines use lo=1, so the levels are inverted by hardware in this circuit
to maintain the portability of the unit between computers. The lines
dio1-4 have not been inverted as the choice of address is arbitrary.)

The presence of an MTA or UNT command (the latter being equivalent to MTA31) on the bus is detected as follows. The MTA code (dio6=0, dio7=1) is converted into a 1 by a NOR gate (IC3) connected to the dio7 and inverted dio6 data lines. The presence of a valid bus command (ie. ATN & DAV both lo) is signalled by a 1 from a NOR gate (IC1) connected to the ATN line of the management bus and the DAV line of the control bus. The output of this NOR gate, ANDed with dio5 to check for address>15, is then ANDed at IC5 with the output of IC3, producing a 1 for MTA>15 commands. This signal operates the follow control of the 4515B decoder, so that the decoder retains the most recent MTA address after the MTA command has been removed from the bus.

Fifteen decoded address lines are available from the 4515B to provide ADC conversion signals and reset signals, etc. and to enable the outputs of several latches holding data. The same output from IC5 also provides a "talk requested" (lo) signal which is stored in a "hold until reset" circuit made up of two NOR gates (IC10 and 11), and NORed with the ATN line (IC12), so that all address lines are made hi as soon as ATN goes lo. The selected address line goes lo only when ATN has returned hi after the MTA command has been decoded. The hold-until-reset circuit ensures that the talk enabled line remains high after any MTA has been received with an address>15, until it is reset by the decoding of an UNT (untalk) command, interpretted using address 15 (from data lines 0-3) and a 1 on d4 (ie. address 31).

The part of this circuit considered so far produces up to fifteen lines which may be activated (lo with a 4515B) by addressing the unit as a Talker and deactivated by untalk. IC4 and IC6 together detect the presence of MLA and UNL codes on the data bus and, using a second hold-until-reset circuit and a second 4515B, produce up to fifteen lines activated by addressing the unit as a Listener and deactivated by unlisten. We have not considered the handshaking control bus in connection with decoding the addresses on the bus because the above circuits can pick up the addresses without any significant delay (when the controller is a relatively slow microcomputer). So provided that all the devices connected to the control bus (including the other parts of this circuit) do handle handshaking correctly, the circuit of fig 7.8 will collect the required addressed without needing specific handshaking circuitry. We shall incorporate handshaking into the data byte handling circuits and these will operate while the address transfers are taking place. So the whole unit will function quite happily whether or not there are other GPIB devices connected to the bus.

At this point it becomes convenient to discuss the remaining elements of the unit in two parts, as one deals with the output of data bytes (from the computer) and the other with data input. The circuit of fig 7.9 shows the handshaking elements of the data output system, addressed as a Listener, and two of the latches used to hold data bytes received over the bus. The address lines come from the

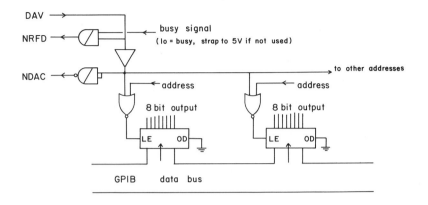

Fig 7.9   The output portion of a multifunction interface useable on a
          GPIB as a listener. This circuit also handles handshaking for
          the address decoding of fig 7.8 by relying on the relatively
          slow response of the computer to NDAC line changes.

listen address 4515B decoder of fig 7.8, and provide a 0 level which
is NORed with the 0 supplied by the DAV line of the control bus. The
circuit releases NRFD (and so allows it to go hi if no other device is
holding it lo) when DAV is hi and when the signal applied to the busy
line is hi. If a busy signal is not required then this line may be
strapped to 5 V. If a busy signal is to be used, derived from, say, a
pulse output device, then it is important that it should not conflict
with the ATN signal provided by the bus controller. As there are no
circumstances (on output) when ATN lo would present a threat to a
device output, a simple solution is to use the ATN line to operate a
digital switch (eg. 4053B) to take the busy line to the busy signal
when ATN is hi, and to 5 V when ATN is lo. When the source device
senses NRFD hi it can load the data bus and set DAV lo. Figure 7.9
responds by setting NRFD lo and NDAC hi, while at the same time
allowing the addressed latch to collect the byte from the data bus.
When the source device senses NDAC hi it returns DAV hi, which results
in the latch enable pin returning lo (preventing any further change in
the output data), NRFD going hi and NDAC going lo. It may be helpful
to follow the signal level changes on the GPIB lines and decoded
address lines in response to a PET PRINT# instruction, and these are
illustrated in fig 7.10.

    Note that the DAV signal is buffered, as it is transmitted to
many other NOR gates attached to the listen latches. This buffering is
unnecessary if only a small number of latches is involved. The SSI
gates in this circuit should be TTL and those with the diagonal line
should be open collector output types to retain compatibility with the
GPIB lines. The latches (or any other circuits used instead) can be
TTL or CMOS, such as the 74LS373 or 74C373 respectively, although if

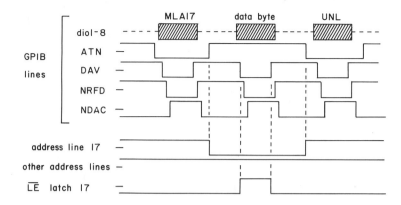

Fig 7.10 Level changes on GPIB data, ATN and handshaking lines and
         interface unit listen address lines during a PRINT#
         transaction.

many TTL devices are to be connected it may be necessary to buffer th
data bus using bus transceivers (see section 7.2.7). It must be
remembered that the data bytes received will be logically inverted
because the GPIB uses lo=1 on the data bus. This can be taken into
account by software (eg. DA = 255-DA before transmission) or by
following the latch outputs with inverters.

     The input section of the interface, addressed as a Talker, is
shown in fig 7.11. The handshaking in this case is initiated by NRFD
going hi, which resets the Q output of a JK flip-flop (the 4027B was
chosen because of the positive transition clock operation) to a 0,
point A on the pulse sketch. Operation of the flip-flop is gated usin
the set (S) input, so that Q may only go lo when the talk address
decoder of fig 7.8 is enabled, ie. when MTA>15 has been decoded and
ATN has returned hi. The Q output signal is passed through an open
collector AND gate (a buffer would do, but there are spare o/c AND
gates on the circuit board from fig 7.9) and supplies the DAV signal.
It also is NORed with the talk address lines to operate one of the
latches and place data on the GPIB data bus. Strictly speaking this
does not follow the GPIB protocol, as data is supposed to be on the
bus before DAV goes lo. However, when a microcomputer is at the other
end it cannot react to DAV rapidly enough to find non-stable data. Th
DAV line remains lo (and the data stays on the bus) until NDAC goes
hi, signifying that the data has been accepted. This transition clock
the flip-flop which, having J=1 & K=0, returns Q hi, which in turn
disables the addressed latch and allows DAV hi (point B on the pulse
sketch).

     The unit described above is effectively limited to the
transmission or reception of one byte per addressing sequence,

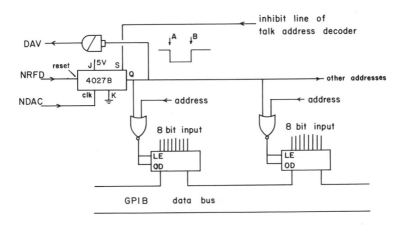

Fig 7.11 The input portion of a multifunction interface useable on a
GPIB addressed as a talker.

although it is not difficult to extend this. For example, the
inclusion of a series of flip-flops, with their outputs NORed with
DAV, in a single address line may be used to read/write several bytes
by sequential handshakes. Several more sophisticated GPIB controlled
interface units are available commercially from companies such as
Small Systems Engineering, Biodata, Machsize and CIL Microsystems. A
versatile MPU based "General Purpose Listener-Talker" is described in
detail in "Interfacing the PET" (see bibliography).

7.2.7 A simple GPIB adaptor

    The circuits described in the last section allow the GPIB
connection of a micro to communicate with a multi-function interface
system for the input and output of bytes, for operating ADCs and DACs
and for switching and sensing signal levels in laboratory equipment. A
quite different problem arises when we wish to attach a microcomputer
to a number of GPIB devices. Unquestionably the best solution is to
use a microcomputer which is fitted with a GPIB compatible I/O port.
Second best is to obtain a high quality commercial micro/GPIB
interface specifically intended for the micro in question, such as the
Interpod system (Oxford Computer Systems (Software) Ltd.) for the CBM
64 and VIC micros. Such interfaces are often constructed around a
microprocessor running software held in ROM, and these are programmed
to handle GPIB transactions correctly. Specific LSI interface circuits
to provide GPIB/MPU signal interfacing are now available for specific
MPUs, and these include the 8292 (Intel) and the TMS9914 (Texas
Instruments). These devices enable microcomputer/GPIB interfaces to be
constructed with relatively little hardware, although requiring the
necessary software. The low-cost alternative described below is only
useable by handling the GPIB management and control bus lines by

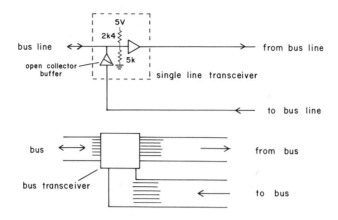

Fig 7.12 A basic transceiver circuit presenting a specified impedance
to the bus line, and asserting the bus line level only when a
lo is required. The lower circuit illustrates how an octal
bus transceiver can split the inputs and outputs from eight
bidirectional bus lines.

software within the microcomputer, and this is only practical using
machine code routines to provide a adequate speed. Even so the speed
is not fast enough to meet all the IEEE 488 timing specifications,
such as the 200ns response to ATN lo. For this reason the GPIB adaptor
must be the only controller on the bus.

First of all we must emphasise that all GPIB lines must be driven
by open collector buffers. (Some lines can be TRI-STATE and although
this allows faster data transfers it is more complex to implement for
a controller, so we will use open collector circuits.) The simplest
way of ensuring that an in house GPIB system meets the required
impedance specifications is to use bus drivers and receivers designed
for this purpose. Typical of these are the bus transceivers (transmitters and receivers combined) MC3446, 3447 and 3448 (Motorola), the
3447 being an octal transceiver and illustrated in fig 7.12. An octal
bus transceiver takes the incoming signals from the bus and reproduces
them on a TTL compatible 8 bit output, and reproduces signals from a
TTL 8 bit input onto the bus. Both the 8 bit inputs and outputs can of
course form buses, but they are unidirectional only and so are
referred to as inputs and outputs, while the bidirectional connections
are referred to as bus lines. Driving bus lines from the input signals
is carried out by open collector buffers with their outputs connected
to both 0 and 5 V through resistances which provide termination
impedances in accordance with the IEEE specification. The buffers
connected to the output lines are not open collector types, but as
their inputs are connected to the same resistances as the bus lines,

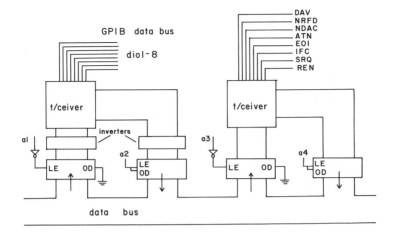

Fig 7.13 A simple micro/GPIB adaptor in which all control and
         management bus lines are software controlled. Note the
         inverters in both data inputs and outputs to handle the
         GPIB's negative logic data.

the absence of any 0 level signal being exerted on the bus
automatically results in TTL hi output levels.

    Two MC3447 bus transceivers form the basis of the GPIB adaptor
shown schematically in fig 7.13. Connection of the adaptor to a micro
could be accomplished using any of the multiplexing techniques
described in chapter 6, but we have only used the synchronous
addressing technique as this allows the adaptor to operate at speeds
comparable with other microcomputer GPIB implementations. Two pairs of
latches are used to handle the input and output bytes, one pair
dealing with bytes for the GPIB data lines and the other pair handling
the control and management bus lines as shown. To simplify the
software the inputs and outputs of the GPIB data bus transceiver are
inverted using 16 inverters, allowing bytes within the computer to be
handled normally. Four address lines (al-4) are used to operate the
latches attached to the MPU data bus, two of these are read addresses
and two write addresses, so only two numerical addresses are
required.

    All control and management bus lines are handled by machine code
routines, and a typical one of these is listed in appendix 3. Actually
the routines are quite straigtforward, if a little tedious, provided
that one remembers to set any outputs hi (logical 0 for data lines,
logical 1 for control and management lines) before attempting to use
them as inputs. (With open collector circuits any one lo output forces
the line lo.)

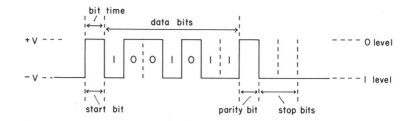

Fig 7.14 The asynchronous serial data format normally associated with
        RS232C systems. The example shown is the character code 105
        (the ASCII character i), even parity with two stop bits. Not
        that a -V level represents logical 1, the first data bit is
        the LSB, and the line idles at the 1 level.

7.3 The RS232C link

An entirely different standard system, primarily intended for
data transfer rather than instrumental control, has become popular fo
the connection of peripherals, such as microcomputers or printers, to
laboratory instruments. This is the RS232 standard of the Electronic
Industry Association, which has undergone two revisions and in its
present form is called the RS232C standard. This often provides a
convenient means of connecting a microcomputer to a commercial
instrument equipped with an RS232C interface, particularly as many
computers (such as the Sirius and the BBC computer) are now fitted
with RS232C-compatible ports, or have RS232C interfaces available as
extras (such as the PET, the Apple and the Spectrum).

7.3.1 Serial data transfers

The particular attraction of the RS232 system is that it is a
serial data transfer system. Rather than having eight wires which eacl
carry the signal level corresponding to one bit of a byte (the
parallel system), a serial system can use as few as two wires (a
signal wire plus ground) to carry the signal level of one bit at a
time. This is probably the most valuable feature of the system in tha
relatively widely spaced units may be connected with minimum amounts
of cabling. The standard specifies allowed speeds (up to 20 kilobits
per second) and signal levels of data transfers (between +/- 5 V and
+/- 15 V, no TTL or CMOS logic levels here), which take place in bit
serial mode.

Although the standard does not specify a particular character
code for the transmission of characters, most modern RS232C systems o
interest here use the asynchronous serial data format, illustrated in
fig 7.14, to transmit a single character of data along the wire (it
need not be 8 bits, so we use the word character rather than byte).

(The RS232C standard also covers synchronous transmissions in which a
clock signal is used, but we shall not be concerned with this mode as
it is rarely used in microcomputer systems.) The transmitted
information consists of a series of signals, each at one of two
allowed levels - the 0 level or the 1 level, (which are usually
voltages such as +6 V for level 0 (the space level) and -6 V for level
1 (the mark level)) - and lasting for a specified period of time.
Before the sequence starts the transmission line is held at level 1,
then a "start bit" of 0 level is sent to indicate that data will
follow. The start bit is followed by up to eight timed applications of
level 0 or 1 making up the data bits as shown in fig 7.14, and then
one (or two) "stop bits"of level 1 to indicate the end of the data and
ensure that the line is held at level 1 before the next start bit is
transmitted.

Clearly the timing of the signals is very important for the
transmission of a series of bits in this manner, and the rate of bit
transmission within a series of bits is called the Baud rate (eg 300
Baud means 300 bits per second). The most commonly used Baud rates are
110 (the standard mechanical Teletype rate), 300, 1200 and 9600. It
should be noted that each 8 bit character transmitted actually
requires 10 or 11 bits to be sent, so that operating at 110 Baud using
11-bit format, the maximum character transfer rate is 10 per second.
Furthermore, the transmission is asynchronous in that start and stop
bits are used to signify the begining and end of each character. The
transmission of the individual bits of a character does depend on
controlled short-term timing, but this timing is important only for
the duration of each character so high precision is not required. One
of the advantages of asynchronous transmission is that characters need
not be contiguous in time, but may be transmitted as they become
available (as for example, from a manually operated keyboard).

It is very important to bear in mind that not all serial
equipment uses the RS232C standard signal levels, even though much
else may be common. There are in particular two important serial
systems which are in widespread use but which can only be connected to
RS232C systems with the aid of a level changing interface. The first
is the (US) MIL standard 188C which differs from RS232C in having the
signal polarities reversed. Generally connecting a 188C device to and
RS232C device will not result in damage to either, although they won't
talk to each other. A more serious problem arise with the variations
of high level interfaces, frequently encountered on old Teletypes
(such as the Model 33) and teleprinter equipment. These interfaces
operated with fairly high voltage levels (typically 60 - 120 V) and
currents adequate to drive mechanical solenoids (20 - 60 mA).
Furthermore, quite a lot of modern serial equipment uses signal levels
compatible with 20 mA current loops, although at lower voltages.
Connecting these systems to RS232C devices can result in damage to the
RS232C circuits. Level changing with a single transistor or op-amp is
quite straightforward. A pair of circuits for connecting a 20 mA loop

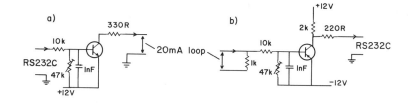

Fig 7.15 Typical circuits used for the conversion of a) RS232C signal
levels to 20 mA loop system, and b) vice versa. Transistors
may be BC477 in a), and BC107 in b).

system to an RS232C device are shown in fig 7.15, although a similar
translation may be achieved using opto-isolators.

Most modern RS232C systems of interest here use 8 data bits to
transmit a single character, although 5, 6 and 7 data bit systems may
also be encountered, and, while any combination of bits may be used,
the most common code for data transmission is the American Standard
Code for Information Interchange (ASCII) shown in appendix 2. The cod
utilises only seven bits, the 128 characters available covering upper
and lower case letters, digits, punctuation marks and about 30 contro
characters such as <carriage return>, <line feed> and <escape>. The
eighth bit is sometimes sent as a 0, but may also be used for parity
checking (which helps to ensure that the other seven bits were
correctly received). Other codes are found on RS232C systems,
including Baudot and Murray codes (5 bit), IBM Correspondence Code (6
bits) and EBCDIC (8 bit).

Computers which are already fitted with an RS232C port are
usually able to transmit or receive data using inbuilt BASIC functions
such as IN, OUT, READ etc. Even where separate interfaces have to be
obtained (as in the case of the PET) these may sometimes be fitted to
the primary I/O port and therefore used easily from within BASIC (eg.
using INPUT#, and PRINT# via the PET's IEEE port). However, there are
a number of variables associated with any RS232C interface, and while
some RS232C interfaces for microcomputers allow the selection of all
relevant parameters, others may be very restrictive. It is essential
to check any interface before purchase to determine what selections
may be made of:

1. Allowed signal levels, +/- 5 V to +/- 15 V or 20 mA loop. Output i
   generally not a problem, as anything above +/- 5 V should be
   suitable. However, the maximum allowed levels for input can cause
   difficulties. The interface should be able to receive +/- 15 V
   without damage.
2. Baud rate, typically covering the range 110-9600 Baud, although
   much broadcast traffic occurs at 40/50 Baud. The "standard" rates

are: 50, 75, 110, 134.5, 150, 300, 600, 1200, 1800, 2400, 3600,
4800, 9600 and 19200)
3. Number of data bits, normally 5, 6, 7 or 8 bits per character.
4. Number of stop bits, should be selectable as 1, 1.5 or 2. (1.5 bits
   means that the stop bit has a duration 1.5 times that of the data
   bits).
5. Parity check or ignore. Both options should be available.
6. Parity even or odd. Again, both options should be available.

The advantages of the RS232C standard interfaces lie primarily in
the ability to use relatively long lengths of connecting cable, which
may consist of as little as two wires for one-way transfer or three
wires for both-way communication, and the relative ease with which
signal levels on such wires may be changed (eg from TTL 0/5 V to +/-12
V, or +/-12 V to 20 mA loop).In principle the lower voltage level
implementations of the RS232C systems are intended for communication
distances of less than 50m, although thousands of metres can be
handled with the 20mA loop system. However, in the laboratory the
limitation of only one device per computer serial port can be severe,
although if the computer's port does follow the RS232C standard there
are actually two communications channels per connector, and with a
certain amount of care it is sometimes possible to make use of both
channels. Serial data transfer can be rather slow if the separate
RS232/computer interfaces have to be used, particularly if the system
is being driven from BASIC or the interface cannot handshake (eg.
several commercial RS232/IEEE interfaces for the PET cannot handle a
data rate of more than 300 Baud under BASIC control). Nevertheless the
existence of an RS232 connection on many larger instruments such as
liquid scintillation counters, multichannel analysers and X-ray
diffractometers, does make the coupling of a computer relatively
simple and, in many cases, a microcomputer, printer and interface may
be cheaper than a new Teletype.

7.3.2 Using RS232C systems with microcomputers

The RS232C standard (which runs to 29 pages) begins with the
words: "This standard is applicable to the interconnection of data
terminal equipment (DTE) and data communications equipment (DCE)
employing serial binary data interchange". Thus fundamental to the
philosophy of the RS232 system is its asymmetry, one device being
regarded as a DCE and the other as a terminal (DTE). The standard
specifies that a 25 pin "D" connector is used for the RS232C link, the
DCE being associated with the female connector and the DTE a male. The
link consists of two channels, a primary and a secondary, both of
which are able to transmit or receive. The names of the signals and
the connector pin numbers are given in fig 7.16. Note that separate
lines are used for transmitted data and received data - the data lines
are not bidirectional. Just to ensure that we all understand the
system used in the RS232C standard note that the data signals (on pins
2 and 3, and 14 and 16) use negative logic, eg. -12 V is a 1, +12 V is

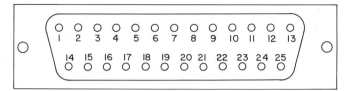

pin no.	direction	name
1	–	protective ground
2	to DCE	transmitted data
3	to DTE	received data
4	to DCE	request-to-send
5	to DTE	clear-to-send
6	to DTE	data set ready
7	–	logic ground
8	to DTE	carrier detected
9		reserved
10		reserved
11		unassigned
12	to DTE	secondary carrier detected
13	to DTE	secondary clear-to-send
14	to DCE	secondary transmitted data
15	to DTE	transmit clock for synchronous data
16	to DTE	secondary received data
17	to DTE	receiver clock for synchronous data
18		unassigned
19	to DCE	secondary request-to-send
20	to DCE	data terminal ready
21	to DTE	signal quality detected
22	to DTE	ring detected
23	to DCE	data rate selector
24	to DCE	transmit clock
25		unassigned

Fig 7.16 The pin connections of the 25 way connector used for RS232C systems. (Rear view of female connector.) Note that some printers now use pin 11 as a handshaking line to indicate "ready for data".

a 0, while the control signals are all positive logic, eg. +12 V on pin 5 tells the DTE that the circuit is clear for it to send data.

The difference between a DCE and DTE is not always as obvious as, say, the difference between a talker and listener on the GPIB, although it may be helpful to remember that the RS232 control lines are named from the point of view of the DTE. Of course the real difficulty is that both the DCE and DTE can transmit and receive characters. The difference between them lies in the use of the many other lines of the 25 wire connection. For example, the DTE can be used to control the levels on the "request to send" and "data terminal ready" lines, while the DCE controls "clear to send", "data set ready", "carrier detected" and "ring detected". Unlike GPIB systems, most of the commonly encountered RS232C systems utilise very few of the available signal lines - although the connector may be hard wired so that, say, "data terminal ready" is always hi (typically +12 V) when connected. In principle the DCE device is that which performs most of the control signal manipulation, and it is often appropriate to regard the microcomputer as the DCE and all attached devices as DTEs. In practice much will depend on the specifications of the other device, ie. which signals it needs or uses, particularly in regard to restricting the rate of character transfer. Certainly the DCE transmits data on pin 3 of its female connector, and receives data on pin 2, while the DTE transmits on pin 2 and receives on pin 3 of its male connector. In most cases commercial equipment with RS232C connections is hardwired and no choice is left to the user.

Microcomputers with built in RS232C ports (which may be primary channel only ports) generally allow the port to be set up for input or for output, and allow characteristics of the serial data (eg. the Baud rate, and sometimes the number of data bits per character, the number of stop bits, and the parity) to be determined by operating system software or a specified array of POKEs. RS232 interfaces used as add-ons for other micros may or may not be so easy to control by computer software, but generally require a more complex arrangement for controlling the transfer of characters - particularly where control is exercised through the same (parallel) port that is handling the characters. Of course it is often possible to avoid character transfer problems by operating at a relatively slow Baud rate, most micros can keep up with a continuous stream of character input at 110 Baud even with BASIC control. However, if a block of multichannel analyser memory is to be transferred frequently into a micro (perhaps 1024 channels, with 6 digits and 2 spaces per channel), then the choice of a low Baud rate can result in a very slow transfer (eg. 20 minutes at 110 Baud with the figures given, compared with about 10 seconds if we could use 9600 Baud!). For character output to, say, a printer the problem is not so serious, as the printer governs that rate at which characters are printed and there is nothing to be gained from trying to send characters to the printer much faster than they can be printed.

If this sounds a little confusing, do not be dismayed. It is confusing. The basic problem is that the RS232 standard does not consider the possibility of handshaking data transfers. On a correct interpretation of the standard the DTE can assert the "request to send" line when it wants a data transfer, and must then wait until the DCE replies by asserting the "clear to send" line before the data transfer can actually take place. However, once the transfer is under way the DCE is not permitted to interrupt it by changing "clear to send" back to a 0. In other words, by asserting "clear to send" the DCE is supposed to be offering unlimited facilities to the DTE until the DTE has finished with it, and not just offering the transfer of a single character - which is the case with normal handshaking systems. A number of device manufacturers have decided to get around this particular difficulty by ignoring the RS232C specification and using "clear to send" (or one of the other lines) as a handshake line anyway. The only difficulty arises when the equipment at the other end of the line does things in a different way. Some RS232 interfaces (eg. the TNW GPIB - RS232 interface) allow the user to address the interface with, for example, one address for characters being input or output and another address for the transfer of control information - such as reading the "request to send" signal or setting the "clear to send" signal, or vice-versa. In these cases the user's program must handle this "throttling" of the character transmission rate. Whenever one is contemplating the purchase of a device claiming an RS232C interface port it is essential to read the detailed specification carefully to understand precisely what each of the connector lines is used for.

The BBC microcomputer contains a serial port related to a newer standard, the RS423. This essentially provides a sub-set of RS232C signals (data in, data out, clear-to-send and request-to-send) so that many RS232C compatible devices may be used when a suitable 5 pin DIN plug/D connector adaptor is made up. The operating system uses the control lines for handshaking, and allows the Baud rates of 75-19200, but not 110.

7.3.3 A simple microcomputer/RS232 adaptor

A simple two way data communication adaptor for converting 8 bit parallel data bus codes into RS232C compatible serial signals is shown in fig 7.17. The brain of the circuit is an LSI IC known as a Universal Asynchronous Receiver and Transmitter, a UART, in this case the 6402, although a host of similar circuits is avaiable. UARTs are particularly useful because they require only a clock signal (which runs at 16 times the required Baud rate of the serial input or output), 8 bit input and output connections to parallel data busses, one serial input and one serial output and a few control signals, to provide both way parallel/serial conversion facilities. Furthermore the format of the serial data, the number of data bits, the number of stop bits and the parity may all be determined by setting logic levels on the control connections, and the UART provides a number of status lines which can be monitored to check when data has been transmitted or received, or whether any error has been detected.

In our circuit the clock signal is supplied to both the received data clock input and the transmitted data clock input (pins 17 and 40 of the UART), but this is not a limitation as the clock rate can be

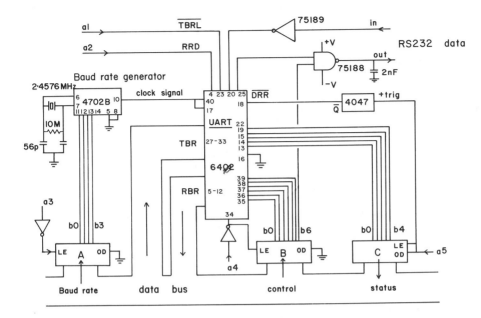

Fig 7.17 A micro/RS232C adaptor based on a 6402 UART.

changed easily by latching into latch A (address a3) a code which sets
the Baud Rate Generator IC, the 4702B, to provide a clock rate
equivalent to any of the standard Baud rates. Byte values of 2-15 (0 &
1 are used for other purposes, see 4702B data sheets) allow Baud rates
of 50, 75, 134.5, 200, 600, 2400, 9600, 4800, 1800, 1200, 2400, 300,
150 and 110 respectively. Similarly codes latched into latch B
(address a4) determine the format of the data according to the
following bit assignments:

b0 parity inhibit (PI); when b0=1 parity generation or checking is
    inhibited. This overrides even parity enable (b4).
b1 stop bit select (SBS); determines the number of stop bits for a
    given number of data bits according to the following rules:

b1 value	character length	stop bits
0	5	1
0	6-8	1
1	5	1.5
1	6-8	2

b2 & 3 character length, ie. number of data bits, according to:

b2 value	0	1	0	1
b3 value	0	0	1	1
no. of data bits	5	6	7	8

b4   even parity enable (EPE); a 1 selects even parity, a 0 selects
     odd parity. Even parity requires an even number of 1s in the
     data+parity bit sequence. This code is only operational when
     b0=0.
b5   master reset (MR); a 1 clears all status bits. Used to reset the
     status bits after an error. Note that this bit must return to a 0
     before normal operation is resumed.
b6   0 selects "break" output (a sustained 0 level) on the serial data
     transmission line. Should be normally 1. Note: not connected to
     UART.

Latch C (address a5) allows a number of status bits to be read by the
computer. The bits have the following meanings:

b0   PE, a 1 indicates a parity error, ie. a received character did
     not conform to the parity specified by b4 of latch B.
b1   FE, a 1 indicates a framing error, ie. the received level was not
     a 1 when a stop bit was expected.
b2   OE, a 1 indicates an overrun error, ie. data has been received
     before a previous character has been transferred to the receiver
     buffer register and DR (b3) reset to 0.
b3   DR, data received flag. A 1 indicates that a data character has
     been correctly received and is ready in the receiver buffer
     register to be placed on the parallel data bus. This bit must be
     reset to 0 (by a lo on DRR) before a new character can be
     received without an overrun error.
b4   TBRE, transmitter buffer register empty flag. A 1 indictaes that
     the transmitter buffer register is empty and ready to receive new
     contents.

Note that addressing latch C also triggers the 4047B monostable, which
pulses the DRR (data received reset) line of the UART. This clears the
data received flag (DR) so that a new character can be received in the
data receiver register. Latch C must be addressed before each
character is "read" by addressing the adaptor with address a2. In
operation for receiving data latch C is examined repeatedly until a 1
at b3 indicates that data is ready for reading from the receiver
buffer register. Address a2 enables the output of the receiver buffer
register and causes the buffer contents to be loaded on to the
parallel data bus.

Data for serial transmission is latched into a transmitter buffer
register from the data bus by addressing the adaptor with address a1.
When this address line returns hi transmission of the character
begins. Note that the UART uses conventional 0 and 5 V logic levels
for all signals. Transmitted data (leaving the UART from pin 25)
passes through an RS232C driver (the 75188 NAND-type gate) where it is
converted into +/-V levels (with -V = 1 logic). This IC therefore uses
supply levels which are different from the others in the circuit and
which can range from +/-5 V to +/-15 V. Received data is converted

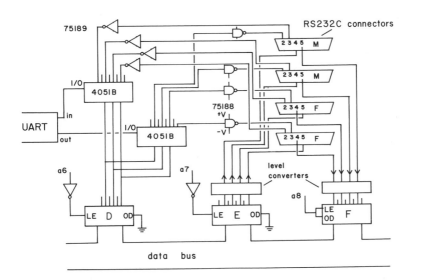

Fig 7.18 An RS232C connector selector for the micro/RS232C adaptor.
The control line logic level converter circuits are shown in
fig 7.19.

from RS232C levels to TTL logic levels by passing through the 75189
(inverting) receiver before reaching the UART.

Examination of the data sheet for the 6402 UART will show that
the above circuit could be implemented without the need to provide
latches B and C (as pin 16 provides and output disable point for the
status lines, and pin 34 accepts a signal to latch the levels of the
control lines. However, it is convenient to use the octal latches as
shown because their spare lines (b7 on latch B and b5-7 on latch C)
may be used for the transfer of levels from control pins of the RS232C
connector. (b6 on latch B may also be used for this purpose if the
"break" facility is not required.) If these spare lines are used for
this purpose it is, of course, essential to provide level changing ICs
in each one (not shown in fig 7.17). Typically request-to-send and
clear-to-send signals may be incorporated in this manner.

Of course the RS232C standard is not a bus system, so it only
allows for one device at each end of the connecting cable. However, it
is possible to make one RS232C adaptor with several RS232C connectors
on it, and to make some of the connectors male and others female. One
way of doing this is to reproduce the type of circuit shown in fig
7.17 several times, having each UART communicating with a single
connector. (UARTs are quite inexpensive, in fact the 6402 costs less
than the 4702B Baud rate generator.) An alternative approach is to
direct the UART's serial input to and output from a selected RS232C

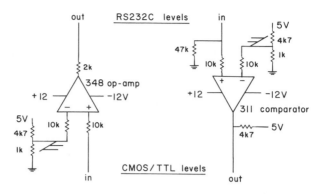

Fig 7.19 The RS232C/CMOS logic level adaptors used for the RS232C
        control lines in fig 7.18. Only one input circuit and one
        output circuit are shown, although 8 are needed in fig 7.18.

connector under computer control as illustrated in fig 7.18.

     This circuit uses a pair of 4051B 8-channel multiplexors (which
allow the signal to pass in either direction) to select which of up to
8 RS232C connectors is connected to the UART. Naturally each input and
output line has an RS232C/TTL level shifting IC included. In the
prototype a second pair of 4051Bs was used to connect to the secondary
channels of each of the 8 RS232C connectors, although these circuits
are not included in fig 7.18. The clear-to-send and request-to-send
lines of each of the connectors were connected through level changing
buffers to the octal latches E and F, allowing the computer to
monitor/control all eight channels while handling data transfers with
just one at a time. Similar arrangements could be used to
monitor/control other RS232C control lines. All incoming control lines
should be biased to ground with a high value resistance to ensure 0
levels at latch F when empty connectors are interrogated.

     The level changing buffers used for the control signal lines are
illustrated in fig 7.19. To produce the +/-12 V control outputs the
0/5 V signals from latch E are used to drive an open loop op-amp to
either of its power supplies, depending on whether its non-inverting
input is less than or greater than the reference voltage (set to about
1 V by a single potential divider). The 348 device used houses four
op-amps in a single package. The incoming RS232C control signals were
converted to CMOS logic levels using 311 comparators, powered at +/-12
V, but with their outputs referenced to ground (pin 1) and connected
through a 4k7 resistor to +5 V. (Note that many single supply quad
comparators, such as the 339, produce the wrong output level if either
input is allowed to go more that about 0.3 V negative.) Both of these
level changes could be accomplished with RS232C compatible drivers and

receivers (eg. 75188 & 75189, both of which house 4 devices per
package), or with opto-isolated couplers.

An additional pair of latches could be used to monitor/control
the secondary clear-to-send and request-to-send lines, although we did
not do this because out of the twenty or so RS232C devices used in our
laboratory, none made any use of these lines. Our prototype RS232C
adaptor had some of its connectors wired for connection to DTEs, and
others for connection to DCEs, hence the asymmetry in fig 7.18.

CHAPTER 8

SYSTEM DESIGN

In the context of the experimental laboratory a microcomputer is just one additional component of laboratory instrumentation, although being very much more versatile than most items of instrumentation because it is programmable. In designing any experimental system a number of logical processes are (or should be) involved and in this chapter we discuss those processes in some detail, because the value of the resulting system depends largely on the care taken in this design stage. Naturally in the present context we are concerned primarily with systems which will involve microcomputers, and certainly an increasing number of laboratory systems are likely to benefit from the use of a micro as one of the component parts. The initial design steps are the same no matter what the system components will be and the choice of a microcomputer component should be made on the same logical grounds as that for any other part of the system. However, once a microcomputer component is decided the design process departs significantly from that used for "hardware-only" systems.

8.1 An approach to system design

Any system is designed by working through a number of individual design steps. The steps follow a logical sequence, although different designers would probably itemise the steps in different ways. One approach suitable for laboratory systems involving a substantial computer component utilises the following discrete steps:

1. Determine system objectives.
2. Assign priorities to system performance.
3. Define system specifications.
4. Determine system components.
5. Outline operational concept.
6. Devise software structures.
7. Design in house hardware components.
8. Decide whether to proceed.
9. Design software overview flowchart.
10. Design software flowcharts.
11. Write software.
12. Test components and assemble system.

Although it is not possible to cover every eventuality, we will discuss briefly each of these steps in turn. To illustrate a realisti design problem we will follow the discussion of each step with an examination of the step taken in one specific design project, the design of a liquid chromatography data collection system. The theoretical basis of this case study need not concern us, although a brief outline of the principal experimental aspects is given below to provide a framework in which the design decisions can be understood. The subject of the case study has been selected because it allows us to cover a moderately wide range of design aspects, although remainin sufficiently simple that it represents a one-man design project which can be discussed in the limited space available.

Case study background

Liquid chromatography is a technique for the separation of small quantities of multicomponent mixtures in solution, usually for the purposes of quantitative analysis. A sample of the mixture is injecte into a stream of liquid (the eluent) flowing into a column packed wit small particles (the column). With a suitable choice of column packin material and eluent it is found that the components of the sample mixture elute from the column after different volumes of eluent have flowed through it, ie. they have been separated from one another. The volume of eluent required to elute a component from the column is known as the component's elution volume. In some cases it is desirabl for the composition of the eluent to change in a controlled manner as elution proceeds; this technique is called gradient elution, and is valuable for increasing the range of components which can be eluted from a single column in a useful period of time. The column eluent is passed into some kind of detector which produces a electrical output related to the concentration of sample component detected. When a suitable detector is used the concentration of each eluted component may be determined from the magnitude of this electrical signal, which is normally recorded on a chart recorder to provide a hard copy "chromatogram". The form of a conventional chromatogram is shown in fig 8.1 and component quantities are often estimated from the peak areas or peak heights recorded in this form.

Conventional chromatographic detectors have a sensitivity switch, and in normal use a sensitivity range is selected before an aliquot o the sample is chromatographed. If an inappropriate selection is made the peaks may be too small or go off scale, so another aliquot is chromatographed to obtain a useable chromatogram. This design project was undertaken because we wished to record useable chromatograms on a routine basis from several hundred samples of widely differing component concentrations as part of a study into ways of increasing the sensitivity of a particular analytical procedure. The chromatographic separations involved were well understood and require a combination of linear gradients which, at a total flow rate of 1 cm^3 min^{-1}, led to complete separations in 15 minutes.

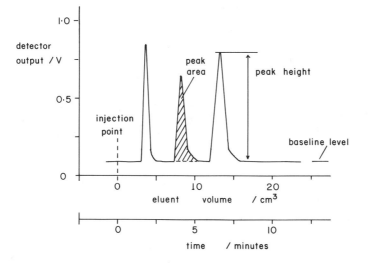

Fig 8.1 A typical chromatogram recorded on a chart recorder with the
0 - 1 V signal from a detector.

Step 1. Determine system objectives

By the time one comes to design a system one probably has a
pretty good idea of what the system will be required to do.
Nevertheless it is useful to discipline oneself to list precisely what
the objectives are for any experiment, and, for systems with several
objectives, to assign a priority to each. For example, if one wishes
to design a system for recording mass spectra then the primary
objective is to obtain the spectra. There may well be other
objectives, such as correcting each spectrum for the efficiency of the
source and the response of the intensity measuring transducer to
different masses, or displaying each spectrum on a video screen, or
saving and recalling spectra using some kind of long term memory
device. However, these objectives probably have a lower priority
because they are irrelevant if the primary objective is not reached.

When deciding on the system objectives it is well worth thinking
ahead to the use one wishes to make of results from the system. It is
relatively easy to build in versatility and expandability to a system
at the begining of the design stage if one has some idea of the
potential areas for later modification. It may be much harder to
squeeze in extra facilities later if no such allowance was made. So,
is there any likelihood that the system will need to compare
experimental spectra with simulated spectra produced using predicted
fragmentation patterns, or that more than one detector may be useful,
or that an X-Y or graphics plotter may be purchased for recording hard
copy?

Case study step 1

    We identified the system design objectives as the following (in approximate order of importance):

1. The system should produce a useable permanent record of gradient elution chromatograms without requiring prior selection of a sensitivity range for the detector.
2. The system should be able to label the chromatograms with retention volumes, peak magnitudes, the areas under specified peaks, and in selected cases the quantity of component detected. The system should have highly flexible facilities for the production of publication-quality graphics hard copy.
3. The system should be able to save and recall spectra, compare two different spectra, subtract one spectrum from another, and deconvolute overlapping peaks.
4. A possible facility to be added later in the project would be a device to enable the automatic injection of sample aliquots, so that many chromatograms could be recorded during overnight, unattended operation.

Step 2. Assign priorities to system performance

    Once the system objectives have been determined it becomes possible to assign priorities to the several aspects of the performance required of the system in reaching the objectives. For example, is speed the most important aspect of performance, or is it ease of use by relatively untrained personnel, or accuracy or precision of recorded data, or the security of data as it is recorded. The priorities assigned at this stage will be referred to in several subsequent steps and, as often the different aspects of performance will be incompatible with one another to some extent (eg. speed and accuracy may lead to different types of interfaces), it is important to have settled in advance which criteria have the highest priority when decisions have to be made.

Case study step 2

    We anticipated that most chromatograms would take between 5 and 20 minutes to collect and that most would be collected by relatively inexperienced presonnel. Amounts available for each sample could be limited, so loss of data through user error should be minimised. These factors determined the ordering of priorities, although we also took into account that there would be several users of the system with various levels of experience. The priority assignments made were:

1. High security of collected data.
2. Ease of use by relatively inexperienced personnel.
3. Automatic shut down in the event of a component failure or inappropriate user action.

Note that high speed dose not figure in our priorities because of the relatively long time scale inherent in the experiments envisaged, neither is accuracy included at this stage. This does not imply that speed or accuracy are unimportant (decisions on these requirements will have to be made in reaching the design specifications), but only that no particular difficulty is envisaged in making the necessary decisions.

Step 3. Define system specifications

At this stage we have decided what the system must do and determined our priorities for the performance of the system. Now it becomes possible to list the design specifications for the system, ie. what signals it will need to accept, what output(s) of information it will need to provide and in what form, what parameters need to be varied and over what ranges, etc. These should be complete specifications (eg. all the outputs required need to be included), as an addition at a later stage could cause major difficulties, but there is no reason why some specifications should not be regarded as optional extras. Generally it is advisable to set both desirable and minimum acceptable specifications, so that there is some room for compromise in step 4 if necessary. For example, it may be desirable that a spectrometer covers the wavelength range 160-900 nm, but the minimum acceptable specification may be a coverage of 200-500 nm. It may be desirable that a system collects the data in 10 ms, and it may or may not be acceptable if 2 hours are needed. The priorities assigned in step 2 will undoubtedly influence the magnitude of the difference between the desirable and acceptable specification figures.

In general one starts with a design specification equal to the desirable specification, altering the design specification if this becomes essential but never going below the minimum acceptable specification. Many factors could force a compromise in specification at several subsequent design stages, even though there may appear to be no reason for such a requirement at this stage. For simple systems which are being designed by one individual all subsequent design steps will be taken by that same individual and modifications to the design specification should be easily adopted. However, for more complex systems where several people may be involved in different aspects of the design it is vital that all should be aware of the design and minimum acceptable specifications, and that any modification of the design specification is quickly notified to all.

Case study step 3

Design specifications connected with the chromatographic system are not relevant in the present context, although decisions on these matters were made at this time. Other design specifications are summarised below (with minimum values in brackets).

Primary input : concentration of reagents in eluent flow, over the
range 0-1000 ppm (parts per million) with resolution better than
0.01 ppm.

Outputs required: graphics output of chromatograms on paper sizes A4
and A3, with resolution of better than 0.1 mm and fully annotated
(sample details, date recorded, axes labelled, peaks annotated
with retention volume and peak height and/or area etc.). Low cost
printer output for listing programs (mainly during development
and testing), numerical data and calculated quantities as
required. Conventional chart recorder output to allow for rough
hard copy. Moderate resolution video graphics for preview of
chromatograms, integration, deconvolution etc., resolution better
than 200*100 pixels.

Controlled parameters: Eluent flow rate variable through range 0-10
(0-5) cm^3 min^{-1}, resolution 0.1 (0.2) cm^3 min^{-1}. Two component
eluent composition variable through range 0-100%, resolution 1%
(2%).

Other specifications:

Status monitoring: Cessation of eluent flow and pump overpressure
conditions should be tested at regular, say 30s, intervals.

Accuracy: Output of chromatogram on graphic plotter or recorder to
better than 1% (2%) of full scale. Calculation accuracy to be
better than 0.1% for all values output, or error estimates to be
included in output.

The use of video and graphics outputs implies digital data handling,
so the rate and duration of data collection must be specified. To
cater for the sharpest peaks likely to be encountered in our
system (known from the column properties) a signal reading rate
of 2 (1) readings per second was determined. The maximum allowed
duration of a single chromatogram was specified to be 25 (20)
minutes, and this is equivalent to 3000 (1500) data readings at
the maximum allowed rate. (Greater flexibility in these
specifications was allowed, but this is largely a chromatographic
matter and will not be amplified here.) The reading closest to
the sample injection point would need to be identified (for
time/volume measurement purposes).

Mass storage requirements: system to have facilities for secure
storage of more than 20 (10) complete chromatograms. Semi-
automatic backup facilities to be included (ie. the system may
instruct the user to perform a backup operation).

Step 4. Determine system components

Now that we have settled on the required inputs, outputs and
operating ranges we can determine what components should be used to
turn the system design specifications into a practical design. This is
probably the most difficult step in the design process, requiring a

good deal of work and a number of answers (both from the designer and
others). Which transducers respond to the property of interest over
the range of values required? What kinds of signal converters are
required to produce the specified outputs from the specified inputs.
Exactly what type of output devices are to be used to present the
required information output(s)? A meter, a digital display, a video
image, a printer hard copy, a conventional chart record or a
multicolour graphics plot? Can some forms of output be provided as
alternatives or as "optional extras" without contravening the design
specification? Does the versatility required or the range of
operations call for a computer and the software writing cost/time that
this would involve, or could an available commercial signal converter
do the job more economically? If a computer is to be used, what would
be our initial guess at a suitable memory size, and would a
microcomputer be adequate or do we need a more powerful minicomputer.
What about mass storage for programs and data? Would a single or dual
disk drive be appropriate, or is the amount of storage likely to be
sufficient to justify a hard disk? Would non-volatile RAM storage be a
good choice for all memory requirements?

Deciding on all these factors can be a laborious business, as at
this stage one needs to read the detailed specification of any
commercial components which may be under consideration as part of the
system. Are the components compatible with one another, and how can
you be sure before you buy? Ancilliary items which seem to be
essential for the required operation of a commercial component must be
noted. The availablility of equipment should be investigated –
particularly recently announced items. Similarly if several items of
computer equipment are contemplated, then the availability of those
items and any relevant software may be different from implications in
glossy advertisements. What kinds of interfaces will be required
between different components? Could any the components which need to
be connected to the computer be obtained with, say, GPIB interfaces?

The software of a computer system is just as much a system
component as any of the hardware items. Who is to provide the
software. You may know BASIC and like the idea of writing the software
yourself. But are you the one who is going to use the system? If not
then its workings may be an inconvenient mystery to whoever is going
to use it – and you will be blamed when it goes wrong and expected to
fix it. Would it be better to talk the user or someone else into
writing the program after all? If speed or security was a high
priority in step 2 then computer programs may need to be compiled. Is
a working compiler actually avaiable for the model of computer you
have in mind? If assembler language software is likely to be needed
who is going to write it? You may know a little assembler, but could
you afford to spend weeks or months writing a massive program that
way, with the high probability that nobody else will ever be able to
modify (or improve) it. And if any special mathematical apparatus is
required in reaching the objectives, is this fully understood –

including the implications of programming it using limited precision
arithmetic routines. To put it bluntly, is the design specification
likely to be jeopardised by the wrong choice of programmer.

At the end of this step you should be able to produce a block
diagram of your system, showing each of the major components and not
forgetting any software component involved. If the systems contains
many components its a good idea to ensure that you can provide for
each block on the diagram details of a) who is supplying the
component, b) the expected delivery time and c) the roughly estimated
cost. In all probability the most difficult part of this will be
estimating the cost (in either money or time) of any software
component of the system, and in practice a reasonable estimate is not
likely to be possible until later in the design process (and a good
estimate probably not until the system has been completed!). Some of
the other estimates will also be vague at this stage (eg. interface
costs), although, as these are not likely to be major costs, they
should provide at least a rough figure to help in deciding whether it
is worth proceeding.

Case study step 4

The sample injection system, column and detector were already
available in our laboratory and need concern us no further, save to
mention that the detector output would be an analog signal in the
range 0-1 V. The number of functions required of the system determine
the use of a computer with associated video graphics display and dual
disk drives (backup requirements dictated that). Peripherals would
need to include a graphics plotter and a cheap printer. The decision
to include a computer allowed us to select an eluent pumping system
which could be controlled by the computer to produce the eluent
composition gradients required. The Applied Chromatography Systems 30(
series pump could pump two liquids independently, and with flow rates
determined by 0-10V analog signals. This offered a solution to both
computer control and gradient elution and, as it was available and
affordable, was selected.

In view of our requirement to record retention volumes with a
precision of better than 2% we also decided to monitor the flow rate
directly, rather than rely on it being determinable precisely from th(
pumping rates set for the two pump heads (the different
compressibility of liquids results in the actual flow rate being
different from the pump displacement rates). We therefore decided to
include a Phase Separations flowmeter to monitor the flow rate of the
eluent after the detector. This flowmeter was available with an RS232(
interface.

A CBM 8032 fitted with high resolution graphics display (200*300
pixels) was selected, partly because the graphics required none of th(

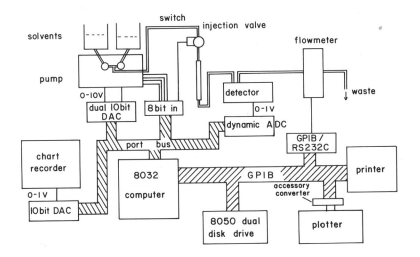

Fig 8.2   Principal system components determined for the chromatography
data system.

normal 32k of RAM; partly because of the ease of ensuring disk
security with the associated 4040 dual disk drive which would also
meet all our requirements for mass storage; partly because the
operating system of the CBM disk unit does not require computer memory
or alter program timings when data has to be written to disk; and
partly because the 80 column display would be suitable for a word
processor. (This was in 1981. There is a wider choice today.) As the
computer was fitted with a GPIB we decided to use a dual channel
GPIB/RS232 interface to allow connection to the flow meter, the second
channel being available for any future requirement at very little
extra cost.

    Having made those decisions most of the other components were
relatively easy to select by reading sales literature. A graphics
plotter and printer with GPIB connectors were selected, to minimise
the pressure on in house interfacing. The block diagram of the system
which developed is shown in fig 8.2, somewhat simplified in respect of
the sample handling end. Some components would need to be constructed
in house, the analog output interfaces for controlling the pumping
rates and providing a conventional chart recorder output, and the wide
range analog input interface for connecting the computer to the
chromatography detector. All other components were available on short
delivery times. The estimated cost of the system seemed justifiable
and the decision to proceed was taken.

Step 5. Outline operational concept

(From here on we will assume that the consideration of
priorities, specifications and system components led us to the
conclusion that a microcomputer should be used as the control and
measurement centre of our system, and that a floppy disk unit should
be used for the storage of programs and data. This decision has a
major effect on the remainder of the design process because the
necessary software is likely to be the principal component not
available "off-the-shelf".)

Having decided what the major components of our system will be we
are now faced with designing one of them, the software, essentially
from scratch. In practice it is this necessity of designing from
scratch which is one of the most interesting aspects of microcomputer
instrumentation design. Designing anything else at this level one
generally has to rely on already existing component parts, with very
little scope for radical improvements in what can be expected from the
end product. However, with software design the situation is quite
different. In most cases there are no already existing components
which have to be used. The designer knows what signal inputs he can
use and what signal outputs his system must provide, and everything
that happens in between can be determined using his own ingenuity.

System software design involves having a concept of what the end
product will look like to the user of the system. Naturally there are
an infinite number of possibilities, but we will confine ourselves to
one relatively pedestrian approach which offers the advantage of being
workable for virtually any system. This is the "selected mode"
approach in which the overall system is capable of operating in one of
a variety of modes, the one required being selected by the system user
with a command from the computer keyboard. Each mode of operation
follows a specific program and is required to perform only a small
number of functions. For example, our chromatography system may be
designed so that in one mode it collects chromatographic data from the
detector and flowmeter, in another mode it displays the chromatogram
on a video screen, in another the chromatographic peaks may be
integrated or deconvoluted, in another one chromatogram may be
compared with another, and so on. Our concept of the end product
system in this case is therefore one of a system centred on the
computer's keyboard and display screen, with the functioning of the
system controlled from the computer keyboard. The modes chosen for any
specific application will, of course, be highly dependent on the
objectives and performance priorities specified earlier.

It is important to think very carefully about the modes of
operation required for the system, both in the light of decisions made
earlier about specifications (such as the form of hard copy output)
and because of the importance of being able to determine at a later
stage how much data to retain in RAM and how many data transactions

between the computer and peripherals or other instruments will be
required. For example, do we wish data collected from a transducer to
be stored on disk immediately, or could these function be performed in
different modes (eg. a collect mode and a save mode)? Is it necessary
to store all transducer readings, or could several be averaged and a
smaller number of averages stored during operation in collect mode.
The answers to these kinds of questions may be important in
determining the maximum rate at which we could take readings under
BASIC control, or in our choice of the interfacing pathway for data
transfer - eg. should the transducer's signal converter and the disk
drive both be on a GPIB?

   Furthermore the software for these various modes of operation may
be separate programs, perhaps written by different individuals.
Defining the way in which data is stored cannot be determined until
one has decided how it needs to be accessed, why it is being stored,
or indeed, what actually needs to be stored. In complex systems the
whole structure of data input and output may be dependent on how many
different functions could be occurring at the same time. For example,
in a system with 12 separate input and output interface channels do we
ever need to communicate with more than 8 functions at once? Obviously
the answer will have a significant bearing on our choice of interface
system.

Case study step 5

   Our operational concept of the chromatography system was very
much of an instrument controlled entirely from the micro's keyboard.
Apart from having to manually inject the sample and occasionally
refill the eluent reservoirs, all aspects of the chromatograph
operation, from collecting the data to producing a fully annotated
diagram of the required size, were to be controlled by the user from
the keyboard. Choosing the modes of operation was not difficult in
this relatively simple case:

 Data collection mode. In this mode the detector output will be moted
   at predetermined times. The data will be stored on disk as it is
   collected, and the pumping system monitored for any failure.
 Chromatogram output mode. The display of chromatograms on the video
   screen would occur in this mode. Any manipulations of the data
   could also be carried out here, correcting the data for changes in
   the detector baseline signal during gradient elution, changing
   units or scales etc. Comparing one chromatoram with another is a
   function which is probably best handled here, as viewing the
   overlapping chromatograms is likely to be the most useful way of
   judging the results. Basically once the user is happy with the
   picture on the video screen, he should be able to instruct the
   system to produce as a hard copy graphics plot a higher quality
   version of the same picture.
 File handling mode. Saving or recalling chromatograms from disk,

creating backup copies of disks, finding out what chromatograms are
stored on a disk, and perhaps editing files (changing a filename or
an erroneous sample detail) could all be handled in this mode.
Automatic sample injection mode. While we did not wish to implement
    automatic sample injection immediately we did expect that at some
    time in the future the system would be modified to collect several
    chromatograms from samples during unattended operation.
Testing mode. One additional mode which would be useful is one which
    allows the detector output to be monitored without data being
    stored, eg. for testing purposes. We would also like to be able to
    program column cleaning flows of solvents, to check the response of
    the flowmeter and detector, and to check the automatic eluent shut
    down facilities which will be incorported to stop the pump in the
    event of a leak. None of these is likely to be of much interest to
    the routine user, but the facilities need to be there just the same
    - how else can the individual parts of the system be tested? These
    facilities can be collected into a testing mode, the existence of
    which need not be advertised to the user.

Step 6. Devise software structures

      Closely related to our decisions on the operating concept for our
system are the requirements which these generate for the software and
storage of data within the system. At this stage it is necessary to
define an outline for the structure of the system software to ensure
that no insuperable problems lie ahead (eg. major bottle-necks in disk
accesses). Is the entire operation to be governed by one program, or
are there to be several - perhaps one for each major mode of
operation? What will each program actually have to do in terms of
inputs and outputs and data handling? Will the software and its
associated data storage fit into the computer's memory space, or
should we allow more memory? Remember most BASICs require at least 5
bytes of RAM for every stored number. Then there the question of
storing data on mass storage. Let us assume that we have already
decided to have disk storage available for saving spectra. What
precisely needs to be stored into disk files? The bytes read in from
transducers, or derived values in floating point form? What else need
to be stored? Do we need to store the wavelength values, the slit
size, the sample identification number, the temperature, the date,
etc. Suppose that one of the modes does provide for correction of
spectra for the efficiency of the optical system, are we going to
store corrected or uncorrected data, or both, and how are we going to
distinguish between the two? Perhaps a few codes at the begining of
each data file would be a good idea, one of these could be used to
indicate whether the contents of the file had been corrected. Another
code might be used to identify which efficiency data had been used to
make the corrections - after all the system's photomultiplier may need
to be replaced after a few month's. Perhaps the correction factors
should be stored in a disk file. Should the data files storing the
spectra be on the same disk as the program(s) or any other data files

or would this slow things down when two different files needed to be read at once.

This stage of the design is more complex than it may appear at first sight, particularly if the experimental system is likely to be producing large amounts of data for several different users. A few 1 Megabyte disks can hold the results of a large amount of work, and changing the file structures after these disks have been filled introduces enormous complexity and aggravation which should be avoidable if sufficient thought goes into the design at an early stage. In this author's view it is wise to build in a small amount of redundant data (such as a few extra records) to file structures at this stage, so that any subsequent minor modifications can at least retain compatability of files.

At the end of this step we should have a firm idea of how many programs are required and how many disk files will be needed. We should also be able to state the maximum amount of memory that any program is likely to need, and specify the maximum number of data values which will need to be handled by the individual programs and stored in the data files. We should be able to draw a block diagram to show which programs will access which data files, and which programs will access which interface units or peripheral devices (printers, plotters, disks etc.). And we should be able to write the guidelines for what each program needs to be able to do.

Case study step 6

The fact that only the data collection and testing modes of operation actually require the control of the pumping and detector systems, and that the software must allow for an expansion at some future date in just this area (the automatic sample injection) suggested that the software should be constructed around two main programs. One of these will handle the apparatus and data collection, and the other the peripherals and chromatogram output, so we will refer to them for the moment as the apparatus and peripherals programs. The ease of use priority determines that transitions from one program to the other must be transparent to the user and require no knowledge of loading programs on the computer. The duration which must be allowed for the longest chromatogram and the time resolution required indicates that up to 3000 readings could be collected in any one chromatogram, and to this must be added up to 100 flowrate values, so we need to allow 16k of RAM for handling this data in BASIC. As this leaves about 15k for software it should not be necessary to divide the programs further, or to compress the data into a smaller space.

Clearly the data needs to be stored in a disk file as it is collected, and this disk file will need to be available to both main programs. This current data file (which we call CURRENT) can be

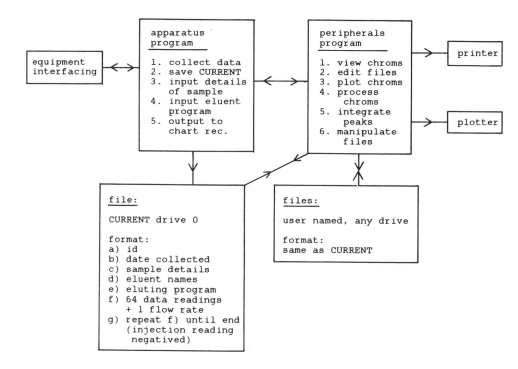

Fig 8.3   Outine of software and data file structure devised for the
          chromatography data system.

provided with all data relevant to the sample and chromatogram during
the operation of the apparatus program, and a copy of this file with
user provided filename could be simply produced without the
peripherals program being involved. The apparatus program will
probably not require access to other data files. The peripherals
program may need to access CURRENT, but it should be possible to
transfer the data between the two programs using RAM. So it will not
be necessary to read in the content of CURRENT every time the program
are swapped. The peripherals program may also need to generate
baseline corrected data, but there is no pressing reason why this
should be filed as it can be quickly regenerated from the content of
CURRENT if necessary. (The correction involves only the subtraction o
data values; no additional quantities are involved.)

     The peripherals program will have facilities for copying
previously stored data files back into CURRENT, for editing the
content of CURRENT and for copying CURRENT into other named files. Th
outline structure devised for the system is illustrated schematically
in fig 8.3.

Step 7. Design in house hardware components

Now that that major aspects of data transfers between the computer and other devices have been considered we may turn to an examination of the detailed requirements of any remaining hardware components. These will be primarily the signal converter/interface components which were not available as suitable commercial units when the system components were selected (or at any rate were not available at the right price). Of course if we have been able to select all the relevant system components to be compatible with one of the standard interface systems, say the GPIB, then this process is relatively straightforward. However, if this has not been the case then now is the time to determine what circuits will be used for these components and how they are to communicate with the micro. Is an interface addressing system required, or does the micro already possess enough I/O ports to handle the individual devices? Do all the devices need to operate at high speed, or can some be more economically implmented using a user port system? If synchronous data transfer techniques are to be used is there a convenient block of addresses free within the micro, or does it allow the use of enough I/O addresses? Will these addresses still be available if the micro has to have its memory size increased to accomodate the software. What is the best source of power for the interface circuits? Could the micro supply this power or should a separate supply be used? If the interface circuits have to be constructed, what is the most economical form of the construction; stripboard, wire wrapping or printed circuit? What are the key points in the circuit for protection against interference, and what connectors and cabling will be required at these points.

At the end of this step we should be able to provide rough circuit diagrams for any hardware components involved, particularly for computer interfaces, and to ensure that the circuits are compatible with one another. This may seem premature, but it is vital to be sure that these small items can be constructed to work together and with the computer. Small errors can be very costly: some TTL circuits may not be available in the 74LS series, and yet a CMOS output cannot drive a battery of standard TTL inputs; some ICs respond to level changes rather than static levels; some interface circuits use or produce hi busy signals, others lo. It should also be possible to estimate the costs of these components more accurately than before. Frequently the cost of cases and connectors will exceed that of the actual ICs involved, although some hybrid ICs are remarkably expensive. The cost of assembly by the chosen method can be estimated adequately from the circuit diagrams.

Case study step 7

Referring back to fig 8.2 we can see that the in house constructed components of the system will be the three analog output interfaces, the high resolution analog input interface, and the safety

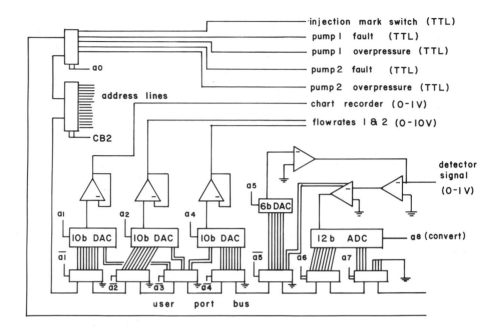

Fig 8.4   The design of the "in-house" interface unit devised for the
          chromatography data system. Only an outline design is given
          here.

sensing unit. The analog input unit will connect directly to
detector's 0-1 V output on the one hand, and some as yet unspecified
part of the computer on the other. No high priority has been placed on
the speed of the data collecting process and it is unlikely that more
than 20 interface transfers will be required in any one second. For
this reason we opt for a multiplexed user port unit of the type
described in chapter 6 (eg. fig 6.10). This unit would contain three
10 bit DACs, one 8 bit logic input system for sensing pump problems
(eg. fig 6.5) and one 12 bit ADC with a variable gain, variable offset
amplifier based on two 8 bit logic output units (one to determine the
gain setting and one for the offset with the aid of an 8 bit DAC, as
illustrated in fig 6.24). These facilities could all be provided
within 16 interface addresses.

     An outline diagram of the interface system in shown in fig 8.4,
where minor components (inverters and gain-setting resistors) have
been omitted. It was decided that the interface system should be
constructed in a single metal case, so only one connection to the
computer's user port would be required. This was taken from the
interface circuit board using an "insulation displacement" plug which
fitted into a standard 14 way DIL IC socket. A coaxial cable fitted
with BNC connectors was used to bring the analog input signal from the

detector to the interface unit, and a BNC connector was fitted to the detector for this purpose. Connection to the detector could have been made via it chart recorder output socket, but we chose to take the signal from the detector's circuit board immediately after its transducer's fixed gain instrumentation amplifier and before it entered the switchable gain and variable offset op-amp which provided the recorder output. Thus the signal we monitored was not effected by front panel controls on the detector, and the recorder output was still available for use. The connections to the pump were made with a six way screened cable terminating at the interface unit end in a 7 pin screened DIN connector.

Step 8. Decide whether to proceed

By now we are equipped with estimates or plans for all components of our system. We can state with reasonable assurance whether the system will meet the objectives, how the performance priorities have been handled and how the system will appear to the user. Somebody now makes a decision about whether to proceed. The parts are ordered, the construction of small circuits undertaken, and we start on the detailed design of the software. However, this is not the point at which the software designer enters the system design process. He has been there at least from step 3. He considered the choice of computer component in step 4. He was also involved in the refinement (or even the creation) of the operating concept in step 5, and should have been a major influence on the choice of program and data structures in step 6. He should also have been closely involved in the selection of interfacing components in step 7, at least to the extent of ensuring that there necessary I/O signals could be handled in the appropriate combinations and in the required timescales. The software is not an afterthought in any computer based system; its one of the major components. Even though the detailed software design has not been started yet, the capabilities of software have already been crucial in deciding what is possible.

Case study step 8

The supplier of the high resolution graphics hardware suddenly announces that the package has been delayed - by three months. We note that his advertisements keep appearing. On the assumption that this means six months delay we plan to improvise with the CBM's low resolution graphics while waiting.

Step 9. Design software overview flowchart

It has already been decided in general terms what functions the software must perform. The task now is to divide these functions into a number of manageable chunks, each of which performs a small number of functions and can be programmed more or less independently. The first stages of software design should involve deciding how to make

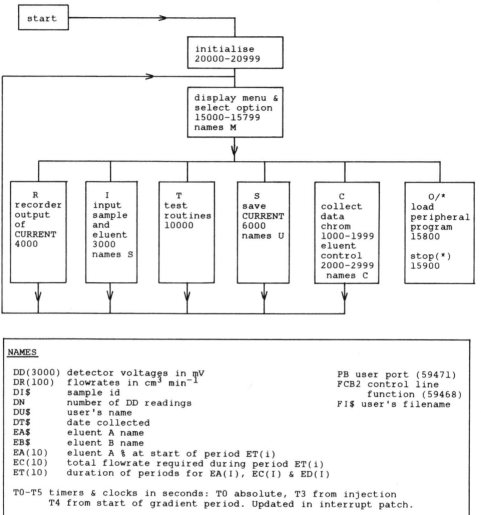

```
NAMES

DD(3000)  detector voltages in mV                PB user port (59471)
DR(100)   flowrates in cm³ min⁻¹                 FCB2 control line
DI$       sample id                                   function (59468)
DN        number of DD readings                  FI$ user's filename
DU$       user's name
DT$       date collected
EA$       eluent A name
EB$       eluent B name
EA(10)    eluent A % at start of period ET(i)
EC(10)    total flowrate required during period ET(i)
ET(10)    duration of periods for EA(I), EC(I) & ED(I)

T0-T5 timers & clocks in seconds: T0 absolute, T3 from injection
       T4 from start of gradient period. Updated in interrupt patch.

temporary variables J,K,L,M,N,O
transfer to compile routines thru CX%(10) & DX%(10)
```

this division and drawing a flowchart which shows how the various
programs and subprograms are linked together. This author's preference
is for menu-driven software, consisting of a master program (sometimes
called a steering segment) which displays on the computer screen a
menu of options which may be selected by the user pressing keys. When
a key is pressed the computer executes the required slave program, and
then returns to the master program and redisplays the menu. In such
cases this first flowchart could have the appearance of fig 8.5,

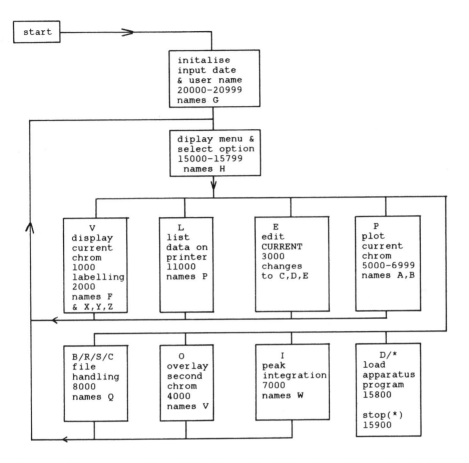

Fig 8.5   Software overview flowchart for the chromatography data
          system. Only representative items have been included here
          owing to limitations on space. Above is the outline of the
          peripherals program. On the previous page is the outline of
          the apparatus program and the principal variable name
          allocaticns.

showing the slave program as boxes. In practice this should be drawn
on a fairly large scale as quite a lot will be written on it as time
goes on. The boxes used will probably bear a resemblance to the
operating modes conceived earlier, although there is no reason why a
one-to-one correspondence should be enforced.

The software overview chart should have certain programming
details added to it as decisions are made; details which will be
essential before the design of the individual slave programs can be
undertaken. For example, which of the slave programs will require

loading from disk, and which will be a subprogram of the master
program (ie. in memory at the same time as the menu displaying
instructions). Those that are subprograms should have blocks of line
numbers or label prefixes allocated, depending on whether the softwa
is to be written in BASIC or assembler. Remember that most BASIC
interpreters search for line numbers from the first line onwards (or
from the current line to the end of program under certain
circumstances - which may not be simply that a destination line numb
is greater than the current line number), so frequently called
routines should have smaller line numbers than rarely used code. In
fact the menu display code could probably have the fairly large line
numbers, and the array dimensioning and data initialisation code the
largest line numbers of all. A block of line numbers (say 2 to 999)
should be set aside for routines which will be frequently called or
fast (such as the interface reading routines), and the RAM area to b
set aside for any machine code routines should be noted.

    In each box should be written a few lines of description of the
function of the program - lines which could be included as REMarks
when the time comes to write the BASIC code. Also written in should t
the names of any disk files which need to be accessed by the program
and, for subprograms, those letters of the alphabet which could be
used for the first letters of names of variables which are local to
that subprogram (this minimises the risk of clashing variable names i
different subprograms).

    In the isolated box (labelled "names" in fig. 8.5) should be
written the names which are to be given to major variables used in
several slave programs, together with a brief note on the quantities
represented (eg. YN(I) = intensity values, WV(I) = wavelength values
and NR = number of readings, etc.). Even when the slave programs are
all independent programs the listings will be much easier to
understand if major variables have the same name in each. It may be
useful for this list to be incorporated in the master program (at the
end, small line numbers are too valuable to waste on many remarks in
BASIC programs). The names to be allowed for short term use variables
should also be included in this box, to minimise the unnecessary
proliferation of unimportant names and waste of space. The names for
small FOR loop variables (eg. II, IJ, IK) and variables used only
during the evaluation of complex expressions (TA, TB) are examples.

    The software overview chart must be completed before the detaile
design of individual programs can be started. Its form must take
account of any relevant priorities assigned to system performance,
such as the security of data - which may require that data is filed t
disk before the user has the opportunity to issue keyboard
instructions and perhaps lose the lot. In this author's experience
several different charts should be drawn so that various options coul
be considered and their merits and weak points weighed before a final
commitment is made.

Case study step 9

An early software overview chart is shown in fig 8.5. Space does not permit the reproduction of a more complete version.

Step 10. Program design

Given a completed software overview flow chart we can now write a detailed specification for each of the programs or major subprograms in turn. For each one we can then draw up a routine flowchart, assigning to each routine a specific role: collecting data from an interface; writing a data file to disk; plotting a array on the screen; listing the data on a printer, and so on. (We use the word routine to mean a small collection of code forming a relatively self-contained part of a subprogram). Each routine can be assigned a group of line numbers, and any routines required by several programs or major subprograms identified and noted on the software overview chart to avoid duplication of effort and wastage of space. Variable names other than the trivial or those already assigned can be chosen using the allowed first letters.

To simplify reading of the programs (particularly in six month's time) suitable REMarks for each program should be included at the begining of each flow chart and at intervals throughout the program. As a matter of good practice programs should have only one exit point (eg. go back to the master program and display the menu) although it is sometimes convenient to allow subprograms to have more than one entry point - preferably with helpful REMarks. This author prefers to attach names to both subprograms and routines. It helps to make the flowcharts easier to read and large programs easier to REMark.

The detailed specifications and flowcharts can now be drawn up for each of the individual subroutines. Each has an assigned collection of line numbers and variables names, and in many cases the flowchart boxes can be completed in BASIC. Any assembler language routines should be flowcharted with as much detail as possible, and labels and names chosen so that the completed program can be checked against the flowchart in the future (after the designer has left or forgotten all about the program).

Case study step 10

Even for this relatively simple design project a considerable number of flowcharts were created. Only two of the subprogram flowcharts are shown to illustrate the procedure. The chart in fig 8.6 is that for the data collection subprogram of the apparatus program and it shows how the interface reading routines are linked in the process of data collection and writing to disk. The chart in fig 8.7 is that for the file handling subprogram of the peripherals program and shows how several different but closely related options selected

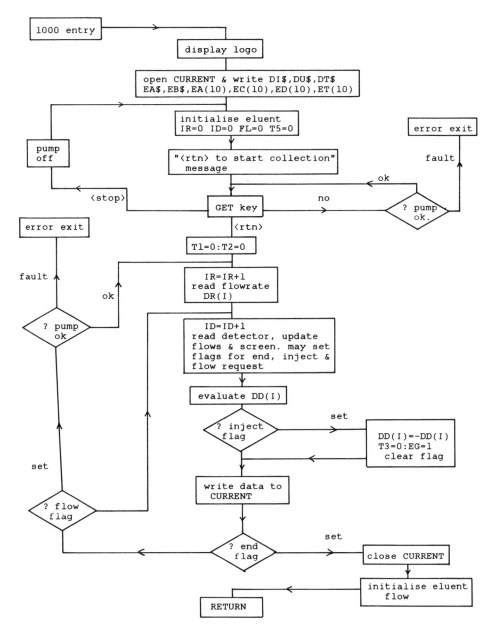

Fig 8.6 Flowchart for the data collection subprogram.

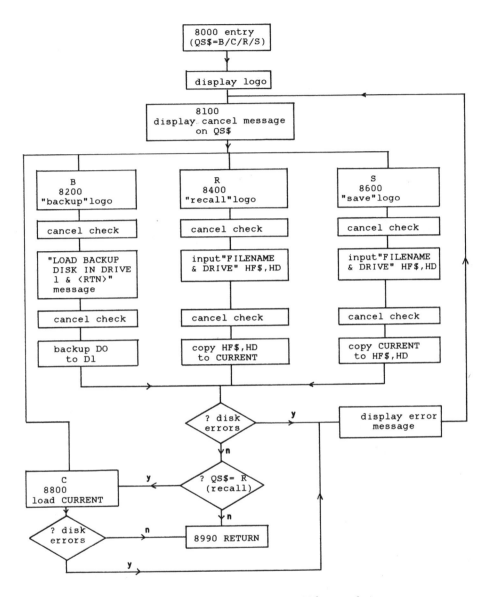

Fig 8.7 Flowchart for file handling subprogram.

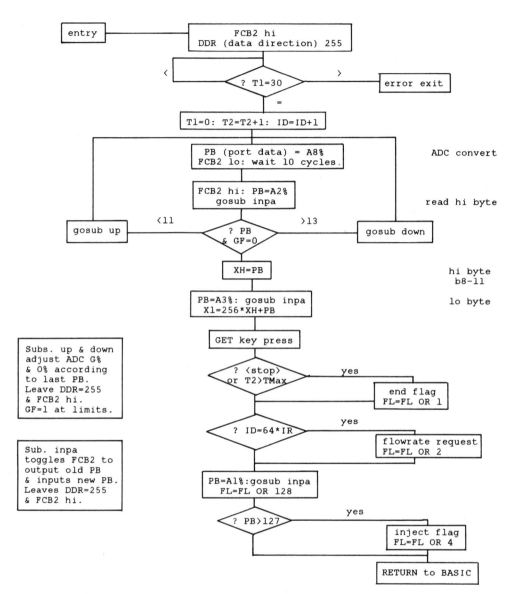

Fig 8.8 Flowchart for detector interface reading routine. Routine
         will be compiled. Parameters initialised with G%(ain)=0
         O%(ffset)=0 and A1%=1: A2%=2 etc.
         Note: transfer back to BASIC is through the 16 bit variable
         X1 & the 8 bit variables G%, O%, and FL% (flags). In PET
         BASIC CX%(1)=X1 CX%(2)=G% CX%(3)=O% CX%(4)=FL% CX%(5)=T1
         CX%(6)=T2 CX%(7)=TM and DX%(i)=Ai%.

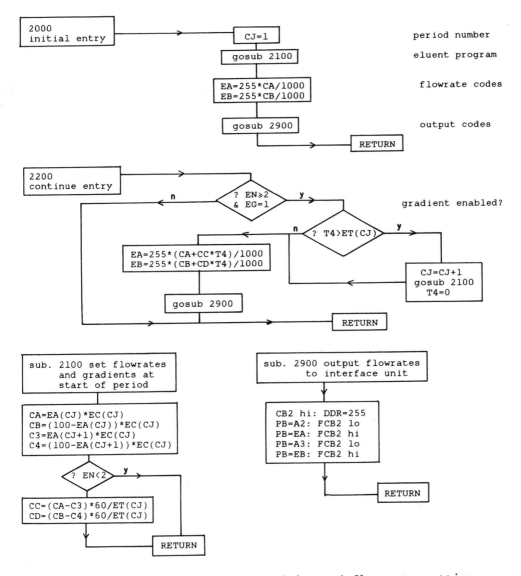

Fig 8.9   Flowchart for eluent composition and flow rate setting
          routine. Parameters provided are T0, T3, T4, EA(i), EC(i),
          ED(i) and ET(i). EG=1 when gradient enabled (ie. after
          injection), 0 when disabled.

from the menu may be handled together to ensure efficient use of
common routines.

Two routine flowcharts are illustrated in figs 8.8 and 8.9. The
first deals with the detector interface reading routine, while the
second is the routine for setting the pump to provide a specified flow
rate of eluent of a specified composition. Both were to be written in
integer BASIC for subsequent compilation, although features of the
language peculiar to integer COMPILED BASIC have been omitted from
these charts.

Step 11. Write software

The routine flowcharts should provide all the information
necessary to enable the programs to be written in the chosen language.
In fact it is only at that stage that one appreciates the effort
devoted to the construction of the flowcharts - or, alternatively,
wishes that more care had been taken in devising them.

Case study step 11

The programs were written in BASIC, with the interfacing routines
compiled using the Integer Basic Compiler supplied by Oxford Computer
Systems (Software) Ltd.

Step 12. Test components and assembled system

Wherever possible individual components should be thorough tested
before the complete system is assembled. In some cases (eg.
interfacing units) the component can be tested only in conjuction with
some of the other components. However, it is nearly always worthwhile
to take some care over component testing, as this inevitably brings
greater familiarity with the individual components. This is
particularly true for software and interfaces, as the nature of
microcomputer suppliers is such that they can rarely provide technical
assistance with laboratory interfacing problems. In the case of
hardware, if the components can be tested individually the results
will often be of assistance if their subsequent interconnection leads
to trouble (through mutual interference, ground loops, etc.).

Once the components are all tested the complete system can be
assembled and tested. Assuming that the hardware components still
work, of major interest at this stage is whether the system meets its
design specifications, whether the software can operate successfully
through many changes of mode, and whether the user interface (ie. the
appearance of the software to the user) is as clear and useful as
planned. Even the best planned systems are likely to require minor
software changes at this stage, so its as well to get them all over
with in one go if possible. However, some aspects of the software
operation will only become irritating or inconvenient as time goes on,

so this author prefers to build up a list of complaints/suggestions over the first week or so of operation by the user(s). At the end of that time a number of software modifications can be incorporated at once, in the full knowledge of the effects such modifications may have on one another. Of course a similar list of suggestion will build up from this time onwards, and it may be desirable to have a reconsider some areas of the software after a few months operation.

Case study step 12

After several faulty components were replaced by the manufacturers the system was used for a period of two months for a preliminary study of the analytical problem in question. At the end of that time it became clear that more space on the data files would be useful to hold additional details about the sample, and that certain aspects of the systems software operation could be improved to prevent user's accidentally deleting their own data files. The latter problem arose because users invented names for their data files, forgot which names they had used (and did not bother to find out), and then saved a second data file with the same name (an intentionally allowed possibility). These problems were rectified and the system has operated ever since.

APPENDIX 1

DECIMAL - HEXADECIMAL CONVERSION TABLES

    The following two pages contain tables for the conversion of decimal numbers into the hexadecimal equivalents. The first table lists the decimal numbers from 1 to 255, ie. the quantities which can be held in a single 8 bit byte and convert to two hexadecimal digits. The second table covers selected decimal values between 256 and 65280, representing the high (most significant) 8 bits of a 16 bit binary number (such as used on an address bus), and converting into the most significant 2 digits of a 4 digit hexadecimal number (the least significant two digits being 00). For any decimal number <32767 the corresponding 4 digit hexadecimal number may be found by determining the nearest lower pair representing the most significant 2 digits, then adding in the least significant 2 digits which correspond to the balance. For decimal number >32767 but <65535 the second table increments in steps of decimal 512.

    In common with most microcomputer literature, we use the $ symbol to represent the hexadecimal notation, eg. $ff02 or $1D00. However, the reader will probably encounter other notations, such as Hffff and 1F1F$_{16}$.

DECIMAL - HEXADECIMAL CONVERSION TABLES

1 $01	2 $02	3 $03	4 $04	5 $05	6 $06
7 $07	8 $08	9 $09	10 $0A	11 $0B	12 $0C
13 $0D	14 $0E	15 $0F	16 $10	17 $11	18 $12
19 $13	20 $14	21 $15	22 $16	23 $17	24 $18
25 $19	26 $1A	27 $1B	28 $1C	29 $1D	30 $1E
31 $1F	32 $20	33 $21	34 $22	35 $23	36 $24
37 $25	38 $26	39 $27	40 $28	41 $29	42 $2A
43 $2B	44 $2C	45 $2D	46 $2E	47 $2F	48 $30
49 $31	50 $32	51 $33	52 $34	53 $35	54 $36
55 $37	56 $38	57 $39	58 $3A	59 $3B	60 $3C
61 $3D	62 $3E	63 $3F	64 $40	65 $41	66 $42
67 $43	68 $44	69 $45	70 $46	71 $47	72 $48
73 $49	74 $4A	75 $4B	76 $4C	77 $4D	78 $4E
79 $4F	80 $50	81 $51	82 $52	83 $53	84 $54
85 $55	86 $56	87 $57	88 $58	89 $59	90 $5A
91 $5B	92 $5C	93 $5D	94 $5E	95 $5F	96 $60
97 $61	98 $62	99 $63	100 $64	101 $65	102 $66
103 $67	104 $68	105 $69	106 $6A	107 $6B	108 $6C
109 $6D	110 $6E	111 $6F	112 $70	113 $71	114 $72
115 $73	116 $74	117 $75	118 $76	119 $77	120 $78
121 $79	122 $7A	123 $7B	124 $7C	125 $7D	126 $7E
127 $7F	128 $80	129 $81	130 $82	131 $83	132 $84
133 $85	134 $86	135 $87	136 $88	137 $89	138 $8A
139 $8B	140 $8C	141 $8D	142 $8E	143 $8F	144 $90
145 $91	146 $92	147 $93	148 $94	149 $95	150 $96
151 $97	152 $98	153 $99	154 $9A	155 $9B	156 $9C
157 $9D	158 $9E	159 $9F	160 $A0	161 $A1	162 $A2
163 $A3	164 $A4	165 $A5	166 $A6	167 $A7	168 $A8
169 $A9	170 $AA	171 $AB	172 $AC	173 $AD	174 $AE
175 $AF	176 $B0	177 $B1	178 $B2	179 $B3	180 $B4
181 $B5	182 $B6	183 $B7	184 $B8	185 $B9	186 $BA
187 $BB	188 $BC	189 $BD	190 $BE	191 $BF	192 $C0
193 $C1	194 $C2	195 $C3	196 $C4	197 $C5	198 $C6
199 $C7	200 $C8	201 $C9	202 $CA	203 $CB	204 $CC
205 $CD	206 $CE	207 $CF	208 $D0	209 $D1	210 $D2
211 $D3	212 $D4	213 $D5	214 $D6	215 $D7	216 $D8
217 $D9	218 $DA	219 $DB	220 $DC	221 $DD	222 $DE
223 $DF	224 $E0	225 $E1	226 $E2	227 $E3	228 $E4
229 $E5	230 $E6	231 $E7	232 $E8	233 $E9	234 $EA
235 $EB	236 $EC	237 $ED	238 $EE	239 $EF	240 $F0
241 $F1	242 $F2	243 $F3	244 $F4	245 $F5	246 $F6
247 $F7	248 $F8	249 $F9	250 $FA	251 $FB	252 $FC
253 $FD	254 $FE	255 $FF			

DECIMAL - HEXADECIMAL CONVERSION TABLES

256	$0100	512	$0200	768	$0300	1024	$0400	1280	$0500
1536	$0600	1792	$0700	2048	$0800	2304	$0900	2560	$0A00
2816	$0B00	3072	$0C00	3328	$0D00	3584	$0E00	3840	$0F00
4096	$1000	4352	$1100	4608	$1200	4864	$1300	5120	$1400
5376	$1500	5632	$1600	5888	$1700	6144	$1800	6400	$1900
6656	$1A00	6912	$1B00	7168	$1C00	7424	$1D00	7680	$1E00
7936	$1F00	8192	$2000	8448	$2100	8704	$2200	8960	$2300
9216	$2400	9472	$2500	9728	$2600	9984	$2700	10240	$2800
10496	$2900	10752	$2A00	11008	$2B00	11264	$2C00	11520	$2D00
11776	$2E00	12032	$2F00	12288	$3000	12544	$3100	12800	$3200
13056	$3300	13312	$3400	13568	$3500	13824	$3600	14080	$3700
14336	$3800	14592	$3900	14848	$3A00	15104	$3B00	15360	$3C00
15616	$3D00	15872	$3E00	16128	$3F00	16384	$4000	16640	$4100
16896	$4200	17152	$4300	17408	$4400	17664	$4500	17920	$4600
18176	$4700	18432	$4800	18688	$4900	18944	$4A00	19200	$4B00
19456	$4C00	19712	$4D00	19968	$4E00	20224	$4F00	20480	$5000
20736	$5100	20992	$5200	21248	$5300	21504	$5400	21760	$5500
22016	$5600	22272	$5700	22528	$5800	22784	$5900	23040	$5A00
23296	$5B00	23552	$5C00	23808	$5D00	24064	$5E00	24320	$5F00
24576	$6000	24832	$6100	25088	$6200	25344	$6300	25600	$6400
25856	$6500	26112	$6600	26368	$6700	26624	$6800	26880	$6900
27136	$6A00	27392	$6B00	27648	$6C00	27904	$6D00	28160	$6E00
28416	$6F00	28672	$7000	28928	$7100	29184	$7200	29440	$7300
29696	$7400	29952	$7500	30208	$7600	30464	$7700	30720	$7800
30976	$7900	31232	$7A00	31488	$7B00	31744	$7C00	32000	$7D00
32256	$7E00	32512	$7F00	32768	$8000				
33024	$8100	33536	$8300	34048	$8500	34560	$8700	35072	$8900
35584	$8B00	36096	$8D00	36608	$8F00	37120	$9100	37632	$9300
38144	$9500	38656	$9700	39168	$9900	39680	$9B00	40192	$9D00
40704	$9F00	41216	$A100	41728	$A300	42240	$A500	42752	$A700
43264	$A900	43776	$AB00	44288	$AD00	44800	$AF00	45312	$B100
45824	$B300	46336	$B500	46848	$B700	47360	$B900	47872	$BB00
48384	$BD00	48896	$BF00	49408	$C100	49920	$C300	50432	$C500
50944	$C700	51456	$C900	51968	$CB00	52480	$CD00	52992	$CF00
53504	$D100	54016	$D300	54528	$D500	55040	$D700	55552	$D900
56064	$DB00	56576	$DD00	57088	$DF00	57600	$E100	58112	$E300
58624	$E500	59136	$E700	59648	$E900	60160	$EB00	60672	$ED00
61184	$EF00	61696	$F100	62208	$F300	62720	$F500	63232	$F700
63744	$F900	64256	$FB00	64768	$FD00	65280	$FF00		

APPENDIX 2

The American Standard Code for Information Interchange
(The ASCII code)

dec	hex	character	name
0	00	NUL	null character
1	01	SOH	start of heading
2	02	STX	start text
3	03	ETX	end text
4	04	EOT	end of transmission
5	05	ENQ	enquiry
6	06	ACK	acknowledge
7	07	BEL	ring bell
8	08	BS	backspace
9	09	HT	horizontal tab
10	0a	LF	line feed
11	0b	VT	vertical tab
12	0c	FF	form feed
13	0d	CR	carriage return
14	0e	SO	shift out
15	0f	SI	shift in
16	10	DLE	data link escape
17	11	DC1	device control #1
18	12	DC2	device control #2
19	13	DC1	device control #3
20	14	DC1	device control #4
21	15	NAK	negative acknowledge
22	16	SYN	synchronous idle
23	17	ETB	end transmission block
24	18	CAN	cancel
25	19	EM	end medium
26	1a	SUB	substitute
27	1b	ESC	escape
28	1c	FS	file separator
29	1d	GS	group separator
30	1e	RS	record separator
31	1f	US	unit separator
32	20	SP	space

The ASCII code

dec	hex	char	dec	hex	char	dec	hex	char
33	21	!	65	41	A	97	61	a
34	22	"	66	42	B	98	62	b
35	23	#	67	43	C	99	63	c
36	24	$	68	44	D	100	64	d
37	25	%	69	45	E	101	65	e
38	26	&	70	46	F	102	66	f
39	27	'	71	47	G	103	67	g
40	28	(	72	48	H	104	68	h
41	29	)	73	49	I	105	69	i
42	2a	*	74	50	J	106	6a	j
43	2b	+	75	51	K	107	6b	k
44	2c	,	76	52	L	108	6c	l
45	2d	-	77	53	M	109	6d	m
46	2e	.	78	54	N	110	6e	n
47	2f	/	79	55	O	111	6f	o
48	30	0	80	56	P	112	70	p
49	31	1	81	57	Q	113	71	q
50	32	2	82	58	R	114	72	r
51	33	3	83	59	S	115	73	s
52	34	4	84	5a	T	116	74	t
53	35	5	85	5b	U	117	75	u
54	36	6	86	5c	V	118	76	v
55	37	7	87	5d	W	119	77	w
56	38	8	88	5e	X	120	78	x
57	39	9	89	5f	Y	121	79	y
58	3a	:	90	60	Z	122	7a	z
59	3b	;	91	61	[	123	7b	{
60	3c	<	92	62	\	124	7c	|
61	3d	=	93	63	]	125	7d	}
62	3e	>	94	64	^	126	7e	~
63	3f	?	95	65		127	7f	del
64	40	@	96	66	‾			

234

APPENDIX 3

Listing of GPOUT routine for communication between PET and GPIB
        adaptor of chapter 7

```
                ;GPOUT
                ;GPIB adaptor byte output routine
                ;also used for addressing when
                ;ATN has been set lo by GPATN.
                ;byte for output enters in acc.
                ;
                al=$a001    ;hardware addresses
                a2=$a002
                a3=$a003
                a4=$a004
                ;
                *=$0360     ;start address
                ;
                ;wait for NRFD hi
nrfdq           bit         a4          ;V set if NRFD hi
                bvc         nrfdq       ;check again if NRFD lo
                ;NRFD hi so load output data
                sta         al
                ;set DAV lo
davlo           lda         #127
                sta         a3
                ;wait for NDAC hi
ndacq           bit         a4          ;N set if NDAC hi
                bpl ndacq ;check again if NDAC lo
                ;reset DAV hi
                lda         #255
                sta         a3          ;handshake complete
                rts
```

BIBLIOGRAPHY

General electronics and integrated circuits

O. Bishop, "Beginner's Guide to Electronics", 4th. edition
   pub. Newnes, London, UK., 1982

G. B. Clayton, "Operational Amplifiers", 2nd edition
   pub. Butterworths, London, UK., 1979

T. D. S. Hamilton, "Handbook of Linear Integrated Electronics for
   Research"
   pub. McGraw-Hill Book Company (UK) Ltd, London, UK., 1977

J. K. Hardy, "High frequency circuit design",
   pub. Reston Publishing Company Inc, Reston, Virginia, 1979

P. Horowitz and W. Hill, "The Art of Electronics",
   pub. Cambridge University Press, Cambridge, UK., 1980

E. Hnatek, "Design of Solid State Power Supplies", 2nd. edition
   pub. Van Nostrand Reinhold, New York, New York, 1981

D. Johnson, L. Johnson and H. Moore, "A Handbook of Active Filters",
   pub. Prentice-Hall, Englewood Cliffs, New Jersey, 1980

W. G. Jung, "IC Op-amp Cookbook", 2nd. edition
   pub. Howard W. Sams & Co., Inc, Indianapolis, Indiana, 1980

D. Lancaster, "TTL Cookbook"
   pub. Howard W. Sams & Co., Inc, Indianapolis, Indiana, 1974

D. Lancaster, "CMOS Cookbook"
   pub. Howard W. Sams & Co., Inc, Indianapolis, Indiana, 1977

G. Loveday, "Essential Electronics, an A to Z guide"
   pub. Pitman, London, UK., 1982

E. S. Oxner, "Power FETs and their Applications"
   pub. Prentice-Hall, Englewood Cliffs, New Jersey, 1982

P. W. Nicholson, "Nuclear Electronics"
    pub. John Wiley & Sons, Inc, New York, New York, 1974

Microcomputers and BASIC

I. Birnbaum "Assembly Language Programming for the BBC Microcomputer"
    pub. The McMillan Press Ltd, London, UK., 1980

R. G. Dromey, "How to Solve it by Computer"
    pub. Prentice-Hall, Englewood Cliffs, New Jersey, 1982

P. Gosling, "Continuing BASIC"
    pub. The McMillan Press Ltd, London, UK., 1980

P. Gosling, "Practical BASIC Programming"
    pub. The McMillan Press Ltd, London, UK., 1980

N. Hampshire, "The PET Revealed", 2nd. edition
    pub. Commodore Business Machines (UK) Ltd, 1980

A. Lee, "PET Machine Language Guide", 3rd. edition
    pub. Abacus Software, Grand Rapids, Michigan, 1981

A. Osborne, "An Introduction to Microcomputers, Vol 1: Basic
    Concepts", 2nd. edition
    pub. Osborne/McGraw-Hill, Berkeley, California, 1979

C. Preston, "A Hitch-Hiker's Guide to the PET"
    pub. ACT (Microsoft) Ltd, Birmingham, UK., 1980

R. West, "Programming the PET/CBM"
    pub. Level Ltd, London, UK., 1982

Microprocessors and assembler language

R. C. Camp, T. A. Smay and C. J. Triska, "Microprocessor Systems
    Engineering",
    pub. Matrix Publishers Inc, Beaverton, Oregon, 1979

R. Bishop, "Basic Microprocessors and the 6800"
    pub. Editions Mengis, 1981

L. Leventhal, "6502 Assembly Language Programming"
    pub. Osborne/McGraw-Hill, Berkeley, California, 1979

L. Leventhal, "6800 Assembly Language Programming"
    pub. Osborne/McGraw-Hill, Berkeley, California, 1979

L. Leventhal, "Z80 Assembly Language Programming"
     pub. Osborne/McGraw-Hill, Berkeley, California, 1979

L. Leventhal, "8080A/8085 Assembly Language Programming"
     pub. Osborne/McGraw-Hill, Berkeley, California, 1979

R. J. Bibbero, "Microprocessors in Instruments and Control"
     pub. John Wiley & Sons, Inc, New York, New York, 1977

Interfacing to digital systems

G. B. Clayton, "Data Converters"
     pub. The McMillan Press Ltd, London, UK., 1982

P. H. Garrett, "Analog I/O Design"
     pub. Prentice-Hall, Englewood Cliffs, New Jersey, 1981

B. A. Artwick, "Microcomputer Interfacing"
     pub. Prentice-Hall, Englewood Cliffs, New Jersey, 1980

J. C. Cluley, "Interfacing to Microprocessors"
     pub. McMillan Publishing Co. Ltd, Basingstoke, UK., 1983

R. Zaks and A. Lesea, "Microprocessor Interfacing Techniques", 3rd.
     edition
     pub. Sybex Inc, Berkeley, California, 1979

J. M. Downey and S. M. Rogers, "PET Interfacing"
     pub. Howard W. Sams & Co., Inc, Indianapolis, Indiana, 1981

E. Fisher and C. W. Jensen, "PET and the IEEE 488 Bus (GPIB)"
     pub. Osborne/McGraw-Hill, Berkeley, California, 1980

Data books

(Most semiconductor manufacturers publish a complete series of data
books detailing the characteristics of their products, and most bring
out updated editions at intervals of only a few years. The books tend
to be collections of data sheets, each individual sheet generally
being available on request from the manufacturer or his local agent.)

eg. National Semiconductor Corporation, Santa Clara, California

Linear Databook (covering op-amps, instrumentation amplifiers, voltage
        comparators, voltage regulators, analog switches, ADCs and DACs,
        transistor and diode arrays)

TTL Databook (covering TTL devices, such as the 74 families, and other
        SSI and MSI circuits not included in the 74 families)

CMOS Databook (covering CMOS devices, such as the 4000 family, the 74C
        family, and other SSI and MSI CMOS devices not included in these
        families)

DEVICE INDEX

SUBJECT INDEX